Reviews of

99 Physiology, Biochemistry and Pharmacology

With 36 Figures

Springer-Verlag Berlin Heidelberg GmbH 1984

ISBN 978-3-662-31034-2   ISBN 978-3-540-38815-9 (eBook)
DOI 10.1007/978-3-540-38815-9

Library of Congress-Catalog-Card Number 74-3674

© by Springer-Verlag Berlin Heidelberg 1984
Originally published by Springer-Verlag Berlin Heidelberg New York in 1984.
Softcover reprint of the hardcover 1st edition 1984

2127/3130-543210

Contents

Indexed in Current Contents

Rev. Physiol. Biochem. Pharmacol., Vol. 99
© by Springer-Verlag 1984

Afferent Vagal C Fibre Innervation of the Lungs and Airways and Its Functional Significance

JOHN C.G. COLERIDGE and HAZEL M. COLERIDGE *

Contents

---

* Cardiovascular Research Institute and Department of Physiology, University of California San Francisco, San Francisco, CA 94143, USA

1 Introduction

The first step towards identifying the impulse traffic in afferent vagal C fibres arising from the lungs and lower airways was taken in the early 1950s by *Paintal* (1955), who observed in cats that injection of phenyl-diguanide into the right atrium evoked low amplitude potentials in small multifibre bundles of the vagus nerve. Comparison of the effects of inject-ing phenyldiguanide upstream and downstream to the pulmonary vascular bed led *Paintal* to conclude that the impulses arose from the lung. The action potentials were of smaller amplitude than those recorded from any other afferent vagal fibre, and the conduction velocities of the fibres were thought to be less than 6 m s^{-1}. Subsequent observations by *Paintal* and others established that these fine fibres were non-myelinated and that they were widespread in the lungs and lower airways in several species (*Paintal* 1964, 1969; *Coleridge* et al. 1965, 1968; *Armstrong* and *Luck* 1974; *Coleridge* and *Coleridge* 1977b; *Russell* and *Trenchard* 1980; *Sapru* et al. 1981).

Long before the impulse traffic in these afferent vagal C fibres was recorded, however, it was widely accepted that a 'fine fibre' afferent vagal input was responsible for initiating the powerful reflex effects observed when irritant gases were introduced into the lower trachea and when certain chemicals were injected into the pulmonary circulation. The most telling observations were those on what came to be called the 'pulmonary chemoreflexes' (a decrease in heart rate and blood pressure, and apnoea followed by rapid shallow breathing) (reviewed by *Dawes* and *Comroe* 1954), which were clearly dependent on afferent vagal pathways, but which could still be evoked when the temperature of the vagus nerves was reduced to 3° or 4°C and conduction in all myelinated fibres eliminated. Moreover, there was evidence to suggest that noxious stimuli such as congestion, embolism and inflammation of the lung exerted at least a part of their reflex effects on breathing and heart rate by engaging this nonmyelinated afferent pathway. Until recently, however, opinion has not generally favoured a reflex role for afferent vagal C fibres under physiological conditions. This is somewhat surprising since more than 40 years ago *Hammouda* and *Wilson* (1939) presented evidence in dogs that small vagal fibres supplied a tonic input to the respiratory centres that increased breathing frequency and decreased tidal volume in the absence of any abnormal stimulus.

The lungs and airways, like the heart and great vessels, have a dual afferent innervation, with an input to the spinal cord as well as to the medulla. The spinal afferents travel in sympathetic nerve branches and are therefore called 'sympathetic afferents' (*Kostreva* et al. 1975). Sympathetic afferents of airway origin are capable of producing disturbances of breathing in response to irritant chemicals (*Widdicombe* 1954c; *Coleridge* et al. 1983), but unlike those that innervate the heart, they do not seem to be involved in the sensation of pain, which is transmitted instead by vagal afferents (*Morton* et al. 1951). We know less about sympathetic afferents from the lungs and airways than about those from the cardiovascular system, and nothing at all about afferent C fibres that may be a component of the former system. Hence our review deals only with the vagal C fibres. Various aspects of the afferent properties and functional role of vagal C fibres arising from the lower respiratory tract are discussed in a number of reviews (*Dawes* and *Comroe* 1954; *Paintal* 1963, 1964, 1973; *Widdicombe* 1964, 1974a, b, 1977a, 1981; *Coleridge* and *Coleridge* 1977c, 1979, 1981, to be published; *Sant'Ambrogio* 1982).

In spite of the gradually accumulating weight of evidence that afferent C fibres from the lungs and airways play a significant role in the neural control of breathing, airway smooth muscle tone, airway secretion, heart rate and peripheral vascular resistance, a general impression persists of a somewhat mysterious afferent system, whose transducer properties are

ill-defined and whose engagement by a variety of foreign chemicals produces only stereotyped and primitive responses of a protective nature.
Perhaps some of the reluctance to accept the hypothesis that non-
myelinated afferents from the lower respiratory tract participate, like their
myelinated counterparts, in regulatory reflexes of a more physiological
nature stems from our general ignorance of the structure and appearance
of the endings themselves. However, a similar lack of information about
the appearance of sensory C fibre terminals in the skin (*Munger* 1971) has
not been an impediment to the general acceptance of the physiological
importance of these cutaneous endings. What is known of the morphology
of the C fibre innervation of the lungs and airways certainly deserves a
place in this account.

2 Morphology

Degeneration studies of the vagus nerve and its branches in cats reveal that
of the 5000 or so afferent fibres distributed to the lungs and lower airways
by each vagus nerve, about 4000 are non-myelinated (*Agostoni* et al.
1957). Nevertheless, the sensory terminals of these non-myelinated afferent fibres have been identified in reasonably large numbers only in the
lungs and intrapulmonary airways of mice (*Hung* ét al. 1972, 1973a, b).
Information about the broad morphological features of this afferent vagal
C fibre innervation, such as the light microscope has provided in the case
of the myelinated afferents supplying the lower respiratory tract (*Larsell*
1921; *Elftman* 1943), is lacking — a deficit that at present shows little sign
of being remedied. For instance, although present evidence suggests that
the sensory terminals of non-myelinated fibres in the lung (*Hung* et al.
1972, 1973a, b) have ultrastructural features in common with the terminals of myelinated fibres (*During* et al. 1974), we do not know whether
the non-myelinated fibres have terminal arborizations.

Electron microscopists attempting to identify non-myelinated afferent
fibres in the lower respiratory tract often confine their attention to the
most distal divisions of the airways, probably in part because *Paintal*
(1955, 1969) suggested that afferent C fibre endings are located in the
alveolar walls close to the pulmonary capillaries, and in part because non-
myelinated fibres in regions of the lung remote from the larger blood vessels and bronchi are thought less likely to be efferent. Using this selective
sampling method, *Meyrick* and *Reid* 1971) found non-myelinated fibres
in the alveolar walls in only 2 of 80 small blocks of lung tissue in 40 rats;
one block contained a single profile of sensory appearance. (A sensory
function is suggested by a terminal axonal enlargement, packed with

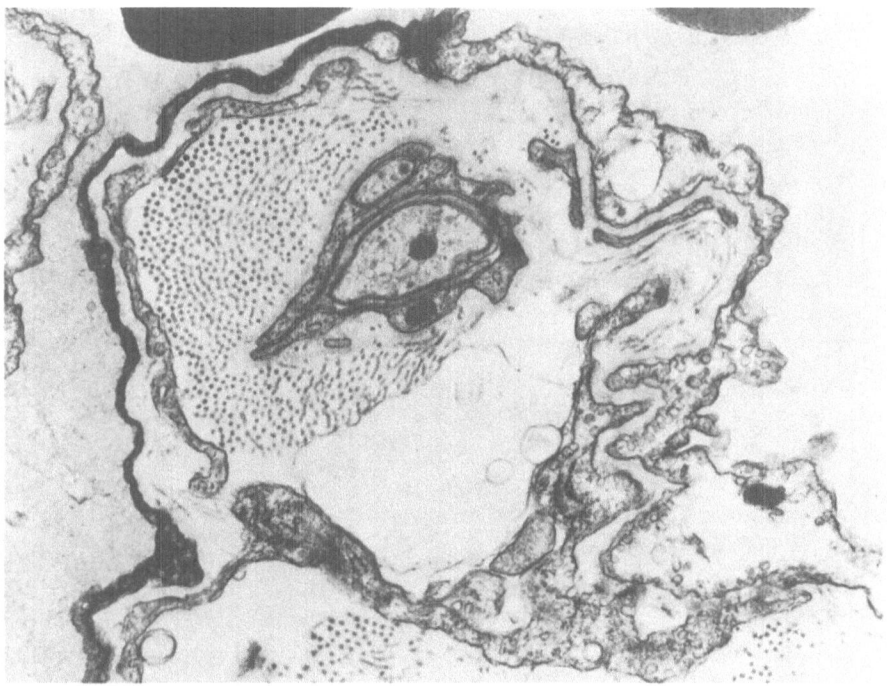

Fig. 1. Innervation of alveolar walls in the human lung. Photomicrograph (× 29100) of tissue taken from periphery of lung just beneath the pleural surface, showing nerve bundle in interstitium of alveolar wall. To *left* and *above* is air space contaminated by red cells. Two non-myelinated axons containing mitochondria, neurofilaments and vesicles are enveloped by Schwann cell cytoplasm and surrounded by collagen fibres. (*Fox* et al. 1980)

mitochondria and only partly ensheathed by a Schwann cell covering, with close apposition between the axonal membrane and some adjacent cell.) In a study of human lung, *Fox* et al. (1980) took samples from the lung periphery immediately beneath the pleura and found non-myelinated fibres in the alveolar walls in 3 of 50 blocks taken from 16 lungs (Fig. 1) but were unable to identify sensory profiles. In studies on both rat and human lung, investigators were impressed by the scarcity of neural elements in the alveolar wall, a finding very different from that in the mouse lung (see below). *Fox* et al. (1980) suggested that the scarcity of nerve fibres in their specimens might reflect a species variation. Such a scarcity may equally be a consequence of the restricted sampling methods, for afferent nerves may be scarce at the periphery of the lung.

Hung et al. (1972, 1973) made a more extensive survey in the mouse and examined serial sections of the entire lung, including the hilar region. They found that non-myelinated fibres were regularly present in alveolar walls and alveolar ducts in all specimens examined. Indeed, several non-

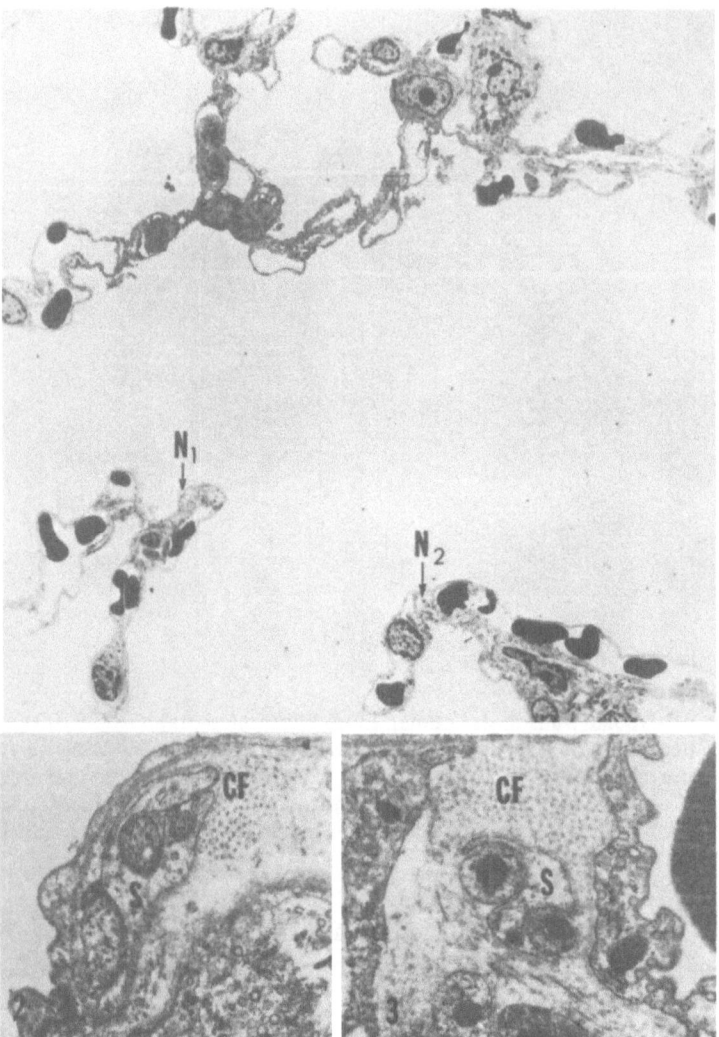

Fig. 2. Innervation of the pulmonary alveoli of the mouse lung. *Above,* electron micrograph of an alveolar duct (× 1710). *Arrows* indicate two bundles of non-myelinated nerve fibres (N_1, N_2) in the interstitium surrounding an alveolar opening. *Below,* higher magnification (× 25 650) of the nerves N_1 and N_2. Three non-myelinated axons can be seen on the *left,* and two on the *right;* the axons are partially or completely surrounded by Schwann cells *(S)* and contain many neurotubules and some mitochondria. *CF,* supporting collagenous fibrils. (*Hung* et al. 1972)

myelinated fibres could be recognized in a single field at lower magnification and were followed through sequential serial sections, at higher magnifications (Fig. 2), to axonal enlargements of a sensory type containing many mitochondria (Fig. 3) and usually associated with type I pneumocytes in the alveolar wall. The sensory enlargements were held to be the

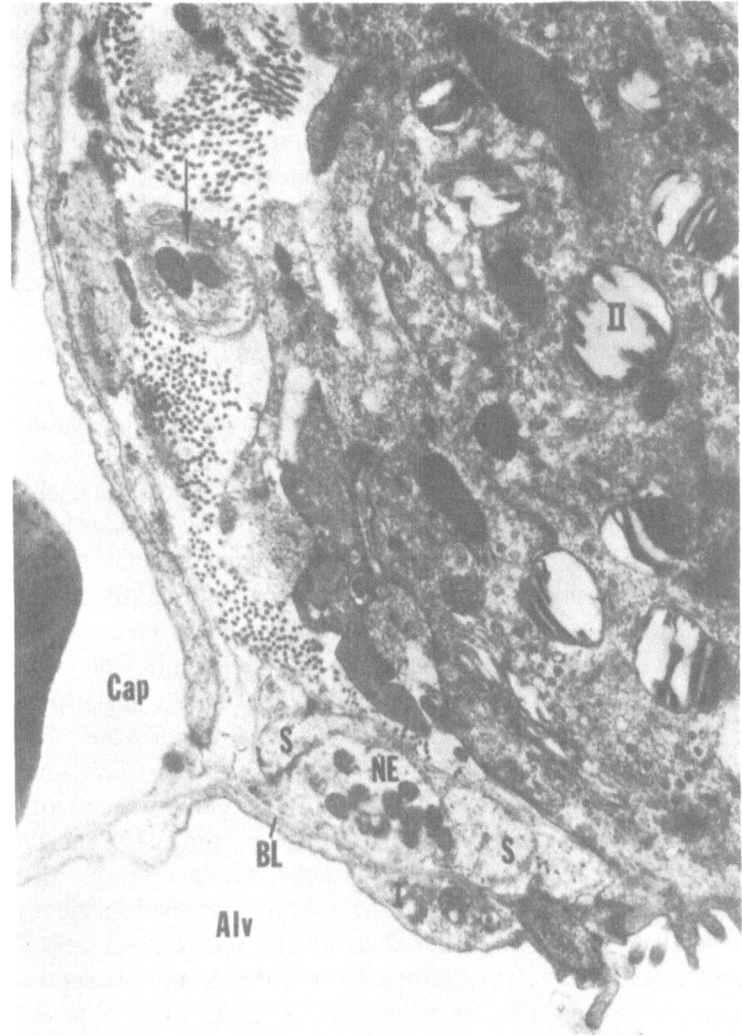

Fig. 3. Innervation of the pulmonary alveoli of the mouse lung. Electron micrograph (× 25 200) showing an enlarged nerve ending *(NE)* in the wall of an alveolar duct *(Alv)*. The ending, which contains numerous small mitochondria, has a bare surface facing the process of a type I pneumocyte *(I)* and separated from it by a basal lamina *(BL)*. The opposite surface of the axon is capped by a Schwann cell sheath *(S)*. A single non-myelinated axon *(arrow)* lies in the vicinity of the ending. *II*, type II pneumocyte. (*Hung* et al. 1972)

equivalent of Paintal's J receptors. Undoubtedly in these studies of mouse lung the investigators greatly increased their chances of finding the elusive non-myelinated fibres by carrying out a broad initial survey at low magnification. To make a comparable survey in human lungs, or even in the smaller lungs of dogs and cats, however, would be a daunting task.

A characteristic feature in rat, mouse and human lung was that most of the axons or axon bundles within the interstitial tissue were surrounded by collagen fibres (Figs. 1, 2). *Paintal* (1970) postulated that J receptors might be surrounded by collagen fibres, the collagen providing an ideal matrix that in the presence of the increased interstitial fluid of pulmonary oedema would swell and become distorted.

Non-myelinated afferent fibres have also been identified in the airways. *Rhodin* (1966) described naked nerve terminals of afferent appearance between the epithelial cells of the human tracheal mucosa and suggested that they corresponded to the sensory nerve endings of non-myelinated fibres that innervate most epithelia. Bundles of non-myelinated fibres were frequently found in the lamina propria of the mucosa and could often be seen to penetrate the outer layers of the basement membrane; they were thought to give rise to the intra-epithelial endings. Single non-myelinated axons, devoid of Schwann cell sheaths, have also been observed between the epithelial cells of intrapulmonary airways in mice (*Hung* et al. 1973b). Some of the axons had terminal enlargements of sensory appearance containing many mitochondria and were identical in appearance with the structures in the alveolar wall believed to be J receptors. Bundles of non-myelinated axons ensheathed in Schwann cells were found in the lamina propria beneath the epithelium and were probably the parent axons. These non-myelinated axons may well correspond to the 'bronchial C fibres' identified in action potential studies (*Coleridge* and *Coleridge* 1977b). The naked axons and axon enlargements were often closely associated with specialized epithelial cells of unusual appearance, containing many dense-cored vesicles. Such cells, which resemble the type I cells of the carotid and aortic bodies, have been described in the airway mucosa of several mammalian species (*Lauweryns* and *Cokelaere* 1973; *Lauweryns* and *Goddeeris* 1975). Some of the nerve terminals associated with these specialized epithelial cells had the characteristics of sensory endings and appeared to be supplied by non-myelinated fibres. The sensory function of these complexes of epithelial cells and nerve endings (the so-called neuro-epithelial bodies) has aroused much speculation.

There is still a great deal to be learned about the structure at the end of the non-myelinated nerve fibre whose impulses we record: the number of terminal branches possessed by a single axon, the size of the terminal arborization and its relation to adjacent structures, the overall distribution of afferent terminals and the density of the sensory fields. These broader features, as well as the ultrastructural detail, are needed to provide the essential link between structure and function. Indeed, it is quite possible that the specialized transducer regions of the terminals themselves, whether of myelinated or non-myelinated fibres and whether mechanosensitive or chemosensitive, all have the same general appearance under the electron microscope. The properties of mechanoreceptors, for example, may be largely determined by the relation of the terminal arborization as a whole to neighbouring visco-elastic elements and smooth muscle cells.

3 Identification and Nomenclature of Lower Respiratory Tract C Fibres

3.1 Identification of C Fibres in Action Potential Studies

The responses of sensory nerve endings are studied most satisfactorily by recording impulses in single nerve fibres. The technique, which involves the splitting of a small nerve slip into finer and finer filaments until a single active unit is obtained, was used by *Adrian* (1933) to record impulses in myelinated vagal fibres arising from pulmonary stretch receptors in cats. *Iggo* (1958) applied the technique to the study of afferent C fibres in the cat vagus and showed that single unit activity can be recorded with a highly satisfactory signal-to-noise ratio, and often with spike amplitudes as large as those of myelinated fibres. This may seem somewhat surprising since several C fibres are known to run together in bundles, each bundle being wrapped in a single Schwann cell sheath; however, there is a constant interchange of C fibres between bundles, and the process of splitting the nerve strands longitudinally will break many of the C fibres as they pass from bundle to bundle (*Iggo* 1958). An alternative technique of recording potentials extracellularly from C fibre cell bodies in the nodose ganglion has been used to investigate non-myelinated vagal afferents from the lung (*Delpierre* et al. 1981).

   Estimates of fibre diameter based on comparison of spike heights are unreliable, for spike heights vary considerably with local conditions, even when the potentials are recorded from two fibres in the same nerve strand, and the distinction between non-myelinated and myelinated fibres must be based on measurement of conduction velocity (*Iggo* 1958). Gasser's value of 2.5 m s^{-1} for the leading edge of the evoked C fibre compound action potential is usually taken as the approximate upper limit for the conduction velocity of vagal C fibres. Afferent vagal C fibres from the lungs and airways have conduction velocities of 0.8–2.4 m s^{-1} (mean 1.4) in dogs (*Coleridge* and *Coleridge* 1977b) and 0.9–2.1 m s^{-1} (mean 1.3) in cats (*Armstrong* and *Luck* 1974); most of Paintal's J receptors in cats had fibre conduction velocities of less than 2.5 m s^{-1} but some had velocities greater than 5 m s^{-1} (*Paintal* 1969).

   Because the process of isolating a single active C fibre may be extremely time-consuming, investigators are sometimes content with preparations that contain several active C fibres. Owing to the difficulty of distinguishing spike heights and configurations in such multifibre bundles, it may be impossible to determine whether the unit responding to one stimulus is the same unit that responds to another; the difficulty is compounded if the stimuli activate additional, previously silent, fibres. Rate-meters with window discriminators set to count potentials of a particular amplitude are useful for prolonged recordings at slow recording speeds, but they can be misleading unless the nerve filament contains no more than two or three active fibres with markedly different spike heights.

Under control conditions many afferent C fibres have such a sparse and irregular discharge that their presence in vagal filaments is easily over-looked. Hence stimulant chemicals such as phenyldiguanide or capsaicin are routinely injected in electrophysiological studies to excite afferent C fibres and to identify their presence (Figs. 4, 5). Input in afferent C fibres

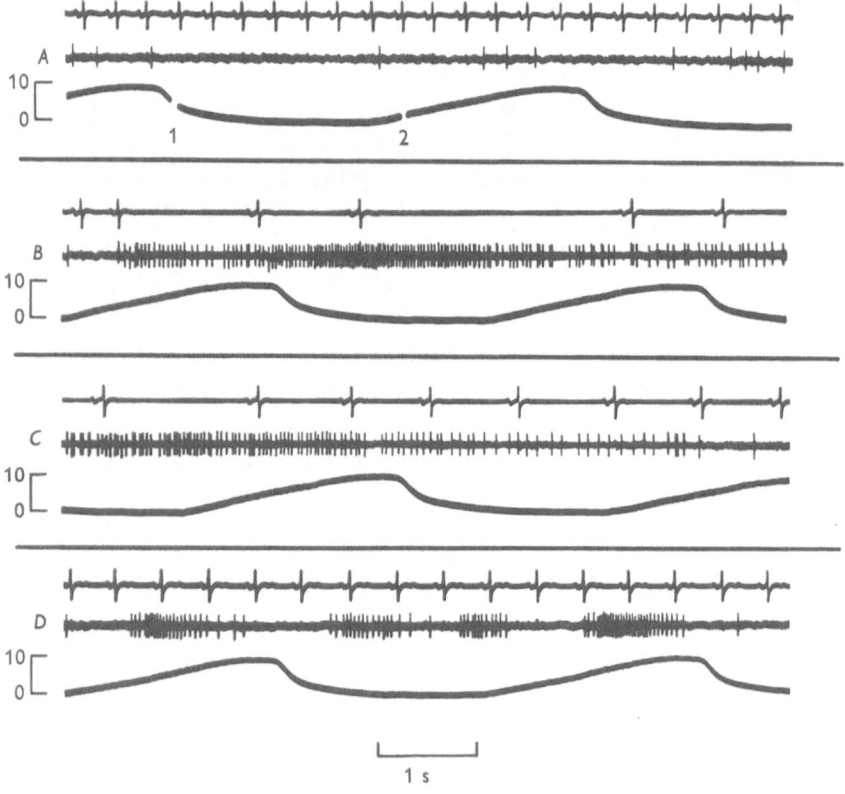

Fig. 4A–D. Stimulation of a pulmonary C fibre in a dog by injection of capsaicin (10 μg kg^{-1}) into a femoral vein (in **A**, between 1 and 2). **A**, **B** and **C** are continuous. Note sparse and irregular discharge in **A**, and abrupt onset of afferent response and cardiac slowing in **B**. (Not shown, left atrial injection of capsaicin had no effect on the C fibre.) **D** Four bursts of impulses evoked when the afferent ending was stimulated by gently pinching the edge of the left lung. The C fibre had a conduction velocity of 0.9 m s^{-1}. *From above downwards:* electrocardiogram; impulses recorded from left vagal filament; tracheal pressure (cm H$_2$O), upstroke representing inflation. (*Coleridge* et al. 1965)

from the abdominal viscera can be eliminated by cutting or ligating the vagus nerves near the diaphragm, but the problem of distinguishing afferent C fibres supplying the respiratory tract from those supplying the heart, great vessels and other thoracic structures remains.

The location of C fibre endings in the lower respiratory tract can be provisionally determined by comparing the afferent responses evoked by injecting stimulant chemicals upstream and downstream to the suspected site of the ending (see below). The precise location of the endings can be determined in artificially ventilated animals with widely opened chests, in which the intrathoracic viscera can be explored and the site of origin of the afferent impulses determined by gently pinching the lung tissue with forceps (Fig. 4C) or by probing the receptive field with a fine rod or bristle (Fig. 6C, D).

3.2 Nomenclature of Lung and Airway C Fibres

In his pioneering studies on the fine afferent vagal input from the lungs in cats, *Paintal* recorded low amplitude activity evoked when phenyldiguanide was injected into the right atrium; the fibres were normally inactive or had a very low rate of discharge (*Paintal* 1955, 1957). The discharge evoked by phenyldiguanide occurred with short latency and coincided with the onset of reflex effects. The endings were supplied by the pulmonary circu- lation and were probably situated near the alveoli; *Paintal* stated unequi- vocally that the endings were not connected with the bronchial circulation or with any part of the bronchopulmonary shunt. Because the endings were also thought to be stimulated by deflation or collapse of the lung, they were called 'deflation receptors' (*Paintal* 1955, 1957, 1963).

The term 'deflation receptor' was very much in keeping with the con- cepts of the vagal control of breathing that were current at the time (*Schmidt* 1941). The central hypothesis, which was based on the studies of *Hering* and *Breuer* (1868), *Head* (1889) and *Adrian* (1933), was that the tonic inspiratory drive of the medullary centres was modulated by two separate and complementary vagal inputs: one set of afferent impulses, aroused by inflation of the lungs, inhibited inspiration; the other, aroused by deflation, excited inspiration. The inhibitory vagal afferents normally kept the medullary centres informed about the volume of the lung and were active during quiet breathing, whereas the excitatory afferents were brought into action when the lungs were deflated and their volume was reduced below functional residual capacity (*Schmidt* 1941). It was also known that tachypnoea could be produced by congestion or embolism of the lung or by inhalation of irritant gases; the tachypnoea was clearly de- pendent upon afferent vagal pathways but it did not appear to involve pulmonary stretch receptors. Since Paintal's studies of the fine afferent input from the lungs were undertaken to identify the mechanism responsible for this tachypnoea, his choice of the term 'deflation receptor' may have been influenced more by the ability of his receptors to evoke tachypnoea, an ability hitherto attributed to hypothetical receptors that were believed to be stimulated by deflation, than by any clear evidence that his receptors were stimulated by deflation. (Indeed, the response of Paintal's endings to deflation was later described as weak and inconsistent, and the name was abandoned.) The term 'deflation receptor' is still occasionally used (*Koller* and *Ferrer* 1973; *Roumy* and *Leitner* 1980), but it now designates rapidly adapting (irritant) receptors with myelinated fibres: such usage adds to the general confusion in regard to nomenclature.

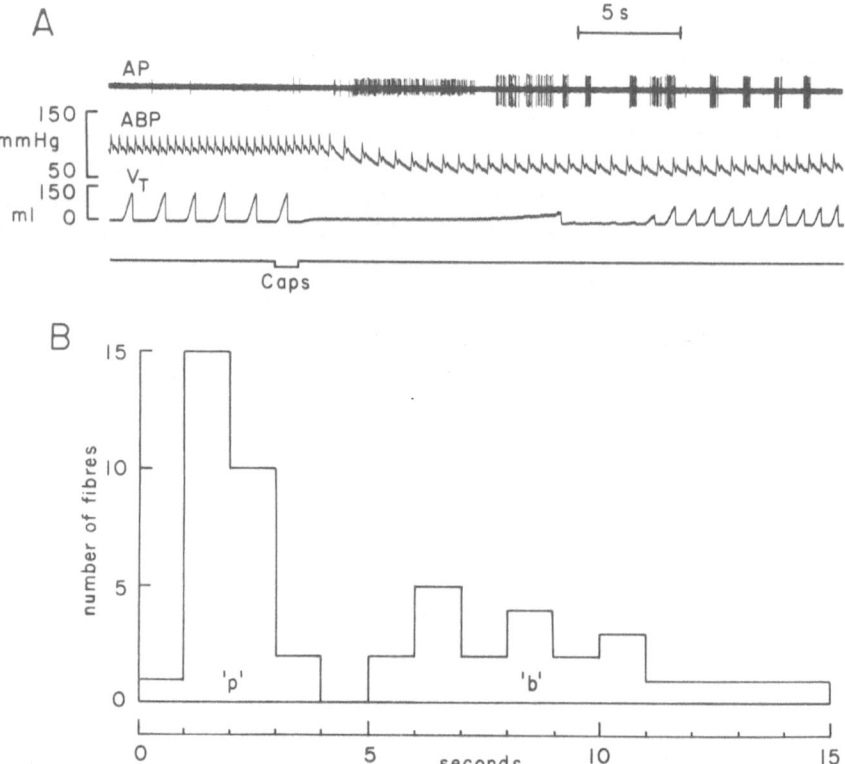

Fig. 5A, B. Comparison of the responses of pulmonary C fibres and bronchial C fibres to injection of 10 μg kg^{-1} capsaicin into the right atrium (at the signal in **A**, at zero time in **B**). **A** Afferent impulses recorded from a vagal filament containing a pulmonary C fibre (small spikes) and a bronchial C fibre (large spikes); both endings were in the right lung; dog breathing spontaneously. *AP*, action potentials; *ABP*, arterial blood pressure; V_T, tidal volume. **B** Histogram showing the latencies of response of 28 pulmonary C fibres ('*p*') and 22 bronchial C fibres ('*b*') to right atrial injection of capsaicin *(Caps)*. (*Coleridge* and *Coleridge* 1977b)

Afferent vagal C fibres were also identified in the lungs of dogs (*Coleridge* et al. 1965, 1968). Their endings were stimulated by capsaicin injected into the pulmonary circulation but were unaffected by phenyldiguanide. Like Paintal's deflation receptors, these C fibre endings had a very sparse and irregular discharge, and they were immediately accessible from the pulmonary circulation but were not reached through the bronchial circulation. Since they were also stimulated by hyperinflation (2−3 V_T) but not at all by deflation or even by collapse of the lung, they were called 'high threshold inflation receptors' (*Coleridge* et al. 1968). The term acknowledged the endings' response to a physiological stimulus within the normal range, and it served to distinguish the C fibre endings from the slowly adapting pulmonary stretch receptors with their generally lower threshold to inflation.

In 1969 *Paintal* re-examined his deflation receptors, confirming that deflation was at best a weak stimulus. By now it seemed clear, both from the immediate afferent response to right atrial injection of phenyldiguanide (coupled with a lack of response to aortic injection) and from the immediate response to insufflation of volatile anaesthetic, that the endings were located in the interstitial tissues close to the pulmonary capillaries. *Paintal* (1969) therefore renamed the endings 'juxta-pulmonary capillary receptors' (type J receptors) − a term that gained wide acceptance and had much to recommend it.

However, we have introduced a new terminology to accommodate our observation in dogs that the endings of many afferent vagal C fibres are located in regions of the lungs supplied by the bronchial circulation (*Coleridge* and *Coleridge* 1977b). The terminology is based upon the vascular accessibility of the afferent endings to stimulant chemicals injected into the bloodstream. The endings of *pulmonary* C fibres are stimulated 0.9−3.3 s (mean 2.1 s) after injection of capsaicin into the right atrium. Pulmonary C fibres are not stimulated at all by injection of capsaicin into the left atrium, nor are they stimulated by injection of small amounts of capsaicin into a bronchial artery (*Coleridge* et al. 1982a). Pulmonary C fibres correspond to high threshold inflation receptors and J receptors, and we think that they are located near the pulmonary capillaries.

The endings of *bronchial* C fibres are stimulated after a long delay by injection of chemicals into the right'atrium, and after a shorter delay by injection into the left. They are also stimulated by injection of small amounts of chemicals into the bronchial artery (*Kaufman* et al. 1980b; *Yamatake* and *Yanaura* 1978). The endings of many bronchial C fibres have been located in intrapulmonary airways, both in large airways near the hilum (Fig. 6) and in smaller airways, some with a diameter of about 1 mm, several centimetres from the hilum (*Coleridge* and *Coleridge* 1977b) .

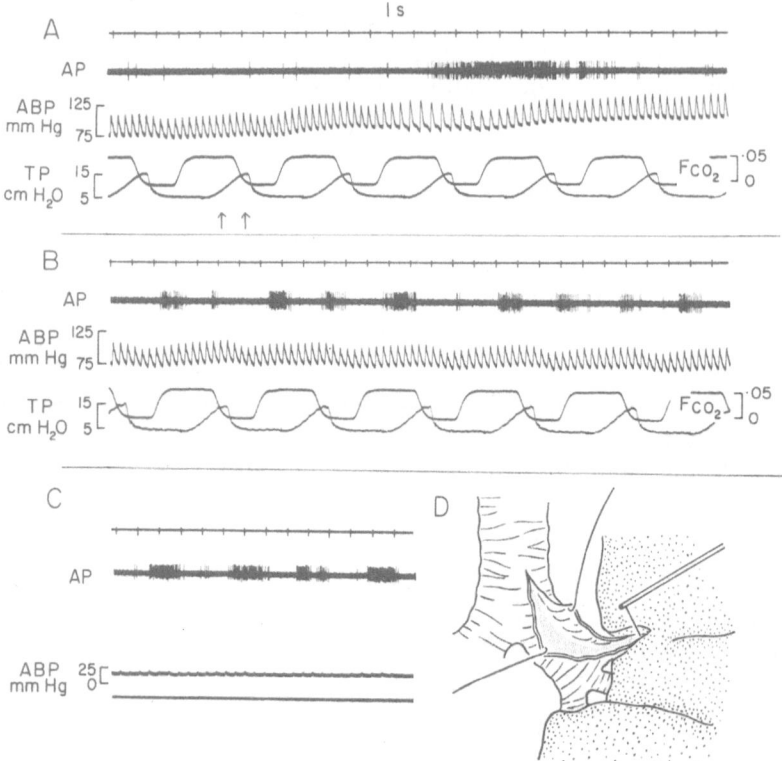

Fig. 6A–D. Stages in the identification of a bronchial C fibre; dog with open chest and lungs ventilated by a pump. In a previous test (not shown) this fibre had been stimulated 9.2 s after phenyldiguanide 10 μg kg^{-1} was injected into the right atrium; the left pulmonary artery was then occluded and remained so for the rest of the experiment. **A** Phenyldiguanide 10 μg kg^{-1} injected into the right atrium *(between the arrows)*. **B** Bursts of firing evoked by repeated probing of the left lung at the hilum. **C** After the dog had been killed, the respiratory pump turned off and the trachea and left bronchus opened; bursts of impulses were evoked by lightly touching the exposed bronchial mucosa with a bristle, as shown in **D**. *AP*, action potentials recorded from left vagal strand; *ABP*, arterial blood pressure; FCO_2, tidal CO_2; *TP*, tracheal pressure. (*Coleridge and Coleridge* 1977b)

Recently impulses have been recorded from afferent vagal C fibres arising from the lower trachea and extrapulmonary bronchi (*Coleridge* et al. 1983). Apart from the location of their endings, these airway C fibres appear similar to the bronchial C fibres arising from the lung.

As we shall see, pulmonary and bronchial C fibres differ not only in their vascular accessibility but also in their afferent susceptibility to a number of mechanical and chemical stimuli. The inference that the intrapulmonary C fibre endings stimulated promptly when chemicals are injected into the right atrium are in the most distal lung divisions (pulmonary C fibres, J receptors) whereas those stimulated at longer latency, and preferentially accessible from the systemic circulation, are in the conducting airways (bronchial C fibres), has been challenged by *Sant'Ambrogio* and *Sant'Ambrogio* (1982) on both theoretical and experimental grounds. These authors point out that the pulmonary and bronchial circulations anastomose widely, and that diffusion distances between the two capillary networks may be very short indeed in some regions of the lung. In experiments in dogs they studied the circulatory accessibility of the intrapulmonary endings of myelinated fibres (slowly adapting pulmonary stretch receptors and rapidly adapting receptors), using veratridine to stimulate endings and the local anaesthetic benzonatate to block them. Although both these types of myelinated fibre are thought to innervate the conducting airways (*Elftman* 1943; *Miserocchi* and *Sant'Ambrogio* 1974; *Mortola* et al. 1975), their responses usually occurred at shorter latency and were more marked when these chemicals were injected into the right atrium than when they were injected into the left (*Sant'Ambrogio* and *Sant'Ambrogio* 1982). It must be stressed, however, that the response to right atrial injection was never so prompt and dramatic as the response of pulmonary C fibres (J receptors) in dogs and cats to right atrial injection of capsaicin or phenyldiguanide, and that the receptors supplied by myelinated fibres were accessible by either route.

If the findings of *Sant'Ambrogio* and *Sant'Ambrogio* should require some modification of the existing view of the location and vascular accessibility of pulmonary C fibres (J receptors), it is not clear in what way. Intrapulmonary endings that are stimulated promptly by chemicals injected into the right atrium but not at all by chemicals injected into the left are preferentially accessible from the pulmonary circulation – and there is no other explanation for their response. The observation can be repeated many times on a given fibre: once a pulmonary C fibre, always a pulmonary C fibre. Moreover, stimulation of a pulmonary C fibre by right atrial injection of capsaicin is abolished by occluding the pulmonary arterial branch supplying the lobe in which the ending is located, and is restored by releasing the lobar branch (*Coleridge* et al. 1965). *Paintal's* hypothesis that the endings have a juxta-pulmonary capillary location is still the most reasonable explanation for the different effects of right and left atrial injection of stimulant chemicals. The morphological studies of *Hung* et al. (1972, 1973a) support this conclusion.

Sant'Ambrogio and *Sant'Ambrogio* (1982) also challenge the hypothesis that the intrapulmonary C fibre endings preferentially accessible from the systemic circulation are probably in the walls of conducting airways. There is good reason to believe that the title 'bronchial C fibre' is justified in the dog, however, for in this species several endings have been located in airways sufficiently large to admit an exploring probe (*Coleridge* and *Coleridge* 1977b), and other C fibre endings have been located in the airways outside the lung (*Coleridge* et al. 1983).

It is conceivable that some afferent vagal C fibres classified as 'bronchial' on the basis of their accessibility to chemicals injected into the bronchial artery were located not in intrapulmonary airways but in the walls of blood vessels whose vasa vasorum stem from the bronchial circulation (*Daly* and *Hebb* 1966), or even in lymph nodes or connective tissue. If such endings have been included among the bronchial C fibres we have examined, their responses to injected chemicals, to inhaled irritant gases or aerosols and to various mechanical stimuli were indistinguishable from those of confirmed airway endings.

In rabbits C fibres with endings in the lungs have been distinguished as 'pulmonary' or 'bronchial' by their latency of response to right atrial injection of phenyldiguanide and by their response or lack of response when injections are made into the left atrium (*Russell* and *Trenchard* 1979). In these smaller mammals intrabronchial exploration is hardly practical; hence the evidence for bronchial C fibres must remain indirect. Some of the observations in rabbits demonstrate that the injection method of localizing intrapulmonary C fibre endings is not ideal; for instance, 3 of 20 intrapulmonary C fibres in rabbits could not be classified as either pulmonary or bronchial, but were equally accessible to injections into either atrium. They were therefore categorized as 'pulmonary–bronchial'. However, the remainder fell clearly into one or other classification, suggesting that some intrapulmonary endings in rabbits, as in dogs, are in conducting airways supplied by the bronchial circulation.

4   Afferent Properties of Lower Respiratory Tract C Fibres

4.1   Response to Chemical Stimuli

Chemicals have played a central role in studies of the impulse traffic in lower airway and lung C fibres, and we begin with a description of the afferent response to chemical stimuli and a brief account of the chemicals that have been most commonly used.

The majority of foreign chemicals and lung autocoids that activate the endings of afferent C fibres in the lower respiratory tract appear to do so directly, rather than by sensitizing them to mechanical stimulation by inflation or deflation of the lungs, by the pulsation of adjacent blood vessels or by the movements of the beating heart. Thus the discharge evoked by chemical stimulation of pulmonary or bronchial C fibres rarely has any obvious relationship to the ventilatory or cardiac cycles and is either continuous (Fig. 4) or irregular (Fig. 5). The irregular discharge of bronchial C fibres frequently takes the form of brief bursts of firing having peak frequencies that may be as high as 50 impulses s^{-1}, the bursts having no obvious ventilatory or cardiac modulation (Figs. 5, 9, 11). Occasionally, however, chemicals appear to sensitize C fibres to the mechanical stimulus of inflation; administration of halothane or other volatile anaesthetics, for example, may cause some pulmonary C fibres to acquire an obvious ventilatory modulation (Fig. 8). We have examined the response of many hundreds of lung C fibres to a wide range of chemical agents, but we have never encountered a cardiac rhythm of discharge. Both slowly adapting and rapidly adapting pulmonary stretch receptors on the other hand often display a conspicuous cardiac modulation after injection of chemicals.

4.1.1 Foreign Chemicals

The foreign chemicals used most commonly in studies of lung afferents were selected in the first instance because they produce reflexes that are thought to originate in the lung. Chemicals used to explore the afferent vagal C fibre system have been further selected for a preferential pharmacological action on the endings of C fibres. They differ in this respect from the veratrum alkaloids, which also initiate reflexes from the lung (*Dawes* and *Comroe* 1954) but do so by a general action on excitable membranes (*Shanes* and *Gershfeld* 1960) which results in the stimulation not only of respiratory C fibres (*Coleridge* et al. 1965) but also of slowly and rapidly adapting pulmonary stretch receptors (*Dawes* et al. 1951; *Sant'Ambrogio* and *Sant'Ambrogio* 1982). Even chemicals that appear to be selective stimulants of C fibres when administered in low doses may have sensitizing or other effects on the endings of myelinated fibres and may even affect the central nervous system itself, when administered in large doses. In his early studies, *Paintal* (1955) used what were probably supramaximal doses of phenyldiguanide and found that they often depressed the response to subsequent injection.

Many chemicals are known to produce pulmonary chemoreflexes by stimulating pulmonary C fibres, but only two, phenyldiguanide and capsaicin, are in common use in both afferent and reflex studies. Both are foreign chemicals with virtually no physiological or pharmacological significance apart from their action on sensory nerves; both cause itch and burning pain when applied to blister base preparations of human skin (*Keele* and *Armstrong* 1964), but they have little else in common.

Phenyldiguanide, an amidine derivative, is thought to owe its action on afferent endings to its structural resemblance to the naturally occurring compound 5-hydroxytryptamine or serotonin (*Fastier* et al. 1959). Serotonin, as well as being a powerful pain-producing agent (*Keele* and *Armstrong* 1964), stimulates afferent vagal C fibres in the intestine and cardiovascular system (*Douglas* and *Ritchie* 1957) and some C fibres in the lower respiratory tract (*Paintal* 1955; *Kaufman* et al. 1980c). Serotonin is also a powerful stimulant of smooth muscle, including, in some species, airway smooth muscle (*Plaut* and *Lichtenstein* 1978). These widespread effects may obscure the primary reflex consequences of stimulation of lung afferents, making serotonin much less suitable than phenyldiguanide for studies of respiratory reflexes. Phenyldiguanide does not appear to share serotonin's powerful direct action on smooth muscle (*Fastier* et al. 1959) although it may have a weak bronchoconstrictor effect in rabbits (*Karczewski* and *Widdicombe* 1969c).

Capsaicin is structurally quite a different compound, a decylenic acid amide of vanillylamine, and the active principle in paprika *(Capsicum annuum);* it causes sensations of tingling and burning when applied to normal skin and is used in folk medicine as a counter-irritant (*Toh* et al. 1955; *Keele* and *Armstrong* 1964). Capsaicin is not known to have any direct effects on smooth muscle. It has been used to study the central connections of chemosensitive C fibres in the somatic nervous system; repeated subcutaneous injection of capsaicin in immature rats in doses $10^4 - 10^5$ times larger than those required to stimulate afferent C fibres causes degeneration of primary sensory neurones in the spinal cord (*Jansco* et al. 1977) and depletes substance P in the substantia gelatinosa of the dorsal horn (*Jessell* et al. 1978), producing analgesia to a wide variety of pain-producing agents.

The general use of two chemicals, rather than one, for identifying pulmonary C fibres in action potential studies and for investigating their reflex functions is a consequence of species differences in the pharmacological properties of this group of non-myelinated afferents. Thus although phenyldiguanide (Fig. 7) and serotonin stimulate pulmonary C fibres and evoke pulmonary chemoreflexes in cats (*Paintal* 1955, 1957, 1969), and phenyldiguanide stimulates pulmonary C fibres and evokes pulmonary chemoreflexes in rabbits (*Karczewski* and *Widdicombe* 1969c; *Russell* and *Trenchard* 1980) and rats (*Sapru* et al. 1981), as a rule neither phenyldiguanide nor serotonin stimulates pulmonary C fibres (*Coleridge* and *Coleridge* 1977b; *Kaufman* et al. 1980c) or evokes pulmonary chemoreflexes (*Dawes* and *Comroe* 1954) in dogs. In the exceptional dog, however, phenyldiguanide injected into the right atrium does evoke the pulmonary chemoreflex; it also stimulates pulmonary C fibres (*Coleridge* and *Coleridge* 1977b). Capsaicin has been the most extensively used of foreign chemicals in studies of pulmonary C fibres and the pulmonary chemoreflex in dogs (*Coleridge* et al. 1964b, 1965, 1982a; *Coleridge* and *Coleridge* 1977b; *Russell* and *Lai-Fook* 1979); it has also been used in cats (*Coleridge* et al. 1968; *Armstrong* and *Luck* 1974) and rats (*Sapru* et al. 1981). In all three species it stimulates pulmonary C fibres (Figs. 4, 5, 7) and evokes the corresponding chemoreflex (Fig. 16).

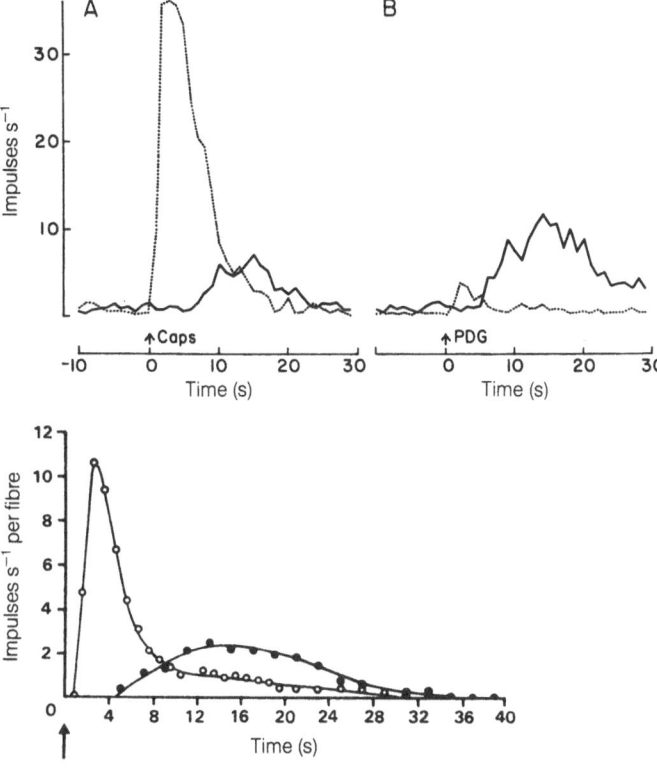

Fig. 7. Effects of chemicals on lung C fibres. *Above,* comparison of the effects of right atrial injection *(arrow)* of 10 μg kg^{-1} capsaicin (**A**) and 10 μg kg^{-1} phenyldiguanide (**B**) on pulmonary and bronchial C fibres in the dog. *Dotted lines,* average impulse activity in 10 pulmonary C fibres; *continuous lines,* average impulse activity in 10 bronchial C fibres, one fibre of each type being examined in each of 10 dogs. *Below,* comparison of the effects of injecting 150 μg phenyldiguanide (○) and 150 μg histamine (●) on impulse activity of type J receptors in the cat; average response of 15 receptors. Chemicals were injected into the right ventricle at zero time *(arrow).* (*Above, Coleridge* and *Coleridge* 1977b; *below, Paintal* 1977b)

In dogs, bronchial C fibres (unlike pulmonary C fibres) are stimulated by phenyldiguanide (Figs. 6, 7) as well as by capsaicin (Figs. 5, 7) (*Coleridge* and *Coleridge* 1977b). This observation provided the first indication that these two categories of afferent C fibre might have different pharmacological characteristics. Phenyldiguanide also stimulates bronchial C fibres in rabbits (*Russell* and *Trenchard* 1980) and cats (*Delpierre* et al. 1981).

C fibres in the lower respiratory tract are also stimulated by chemicals added to the inspired air. Pulmonary C fibres are stimulated by powerful airway irritants such as chlorine (*Paintal* 1969) and ammonia (*Armstrong* and *Luck* 1974). They are also stimulated by high concentrations of the volatile anaesthetics halothane, trichlorethylene, ether and chloroform (*Coleridge* et al. 1968; *Paintal* 1969). In artificially ventilated dogs and

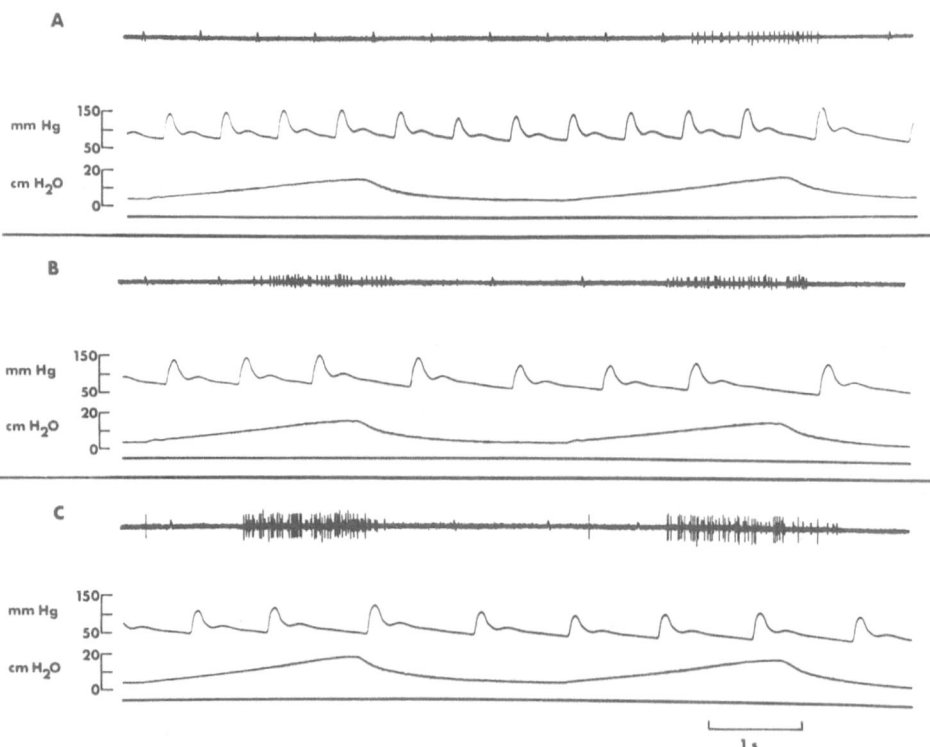

Fig. 8A–C. Stimulation of two pulmonary C fibres (small and large spikes) by halothane in a dog with open chest and lungs ventilated by a pump. **A** 5% halothane had
already been administered for 10 s and was continued throughout **A** and **B**; note that
both fibres were silent in the first ventilatory cycle, as they had been throughout the
preceding control period, but that one fibre became active in the second cycle. **B** Approximately 10 s later. Between **B** and **C**, the concentration of halothane was increased
to 10%. **C** 15 s later; note that the previously inactive fibre was now stimulated (large
spikes). Both endings were located in the left lower lobe. *From above downwards:*
impulses recorded from left vagal strand; arterial blood pressure; tracheal pressure.
(*Coleridge* et al. 1968)

cats the response of a few pulmonary C fibres to these anaesthetics takes
the form of a continuous but irregular discharge without evidence of
ventilatory modulation. The majority of pulmonary C fibres, however,
display a different pattern of response consisting of a somewhat irregular
train or burst of impulses in time with lung inflation (Fig. 8) and having
a maximal frequency of 50–60 impulses s^{-1} (*Coleridge* et al. 1968). Tracheal pressure in these artificially ventilated animals sometimes increased
slightly during administration of the anaesthetic, but the increase in C fibre
activity invariably preceded any increase in pressure by at least one or two
ventilatory cycles. The absence of any obvious change in lung compliance
at the onset of stimulation coupled with the known sensitivity of pulmonary C-fibres in both dogs and cats to large lung inflations (see below)

suggested that the majority of pulmonary C fibres were sensitized by volatile anaesthetics to the effect of inflation (*Coleridge* et al. 1968). Regardless of the pattern of response, firing began within a few ventilatory cycles of the administration of anaesthetic, often during the first or second cycle. *Paintal* (1969) reports that the latency of response of J receptors to insufflation of halothane may be as short as 0.3 s. The stimulation of pulmonary C fibres coincides with the depression or abolition of the activity of slowly adapting pulmonary stretch receptors by the anaesthetic vapour (*Coleridge* et al. 1968).

The response of bronchial C fibres to inhaled irritants has not been examined so extensively. However, both pulmonary and bronchial C fibres, the latter including C fibres supplying the extrapulmonary bronchi and the lower part of the trachea, were stimulated by SO_2 administered to the lower airways in dogs (*Roberts* et al. 1982b). The discharge evoked by SO_2 often waxed and waned in a rhythmical fashion, with a time course that paralleled the vagally mediated reflex effects.

4.1.2 Response to Lung Autocoids

C fibres in the lower airways are not only stimulated by foreign chemicals, they are also stimulated by the administration of lung autocoids, i.e. chemicals that are formed and released in the lungs and airways in a variety of physiological and pathological conditions. The autocoids known to stimulate airway C fibres are histamine, the prostaglandins, bradykinin and serotonin. All but serotonin are released in the lungs of dogs and other mammals in pulmonary anaphylaxis and their release is believed to play a major role in the asthmatic syndrome in man (*Nakano* and *Rogers* 1976; *Garcia Leme* 1978; *Plaut* and *Lichenstein* 1978). Serotonin release contributes to immune-type phenomena in rodents but not in dogs or man; however, serotonin is released from platelets after pulmonary embolism (*Plaut* and *Lichenstein* 1978).

All the above compounds qualify as 'algesic agents' in tests in human subjects and animals (*Keele* and *Armstrong* 1964; *Moncado* et al. 1978) and, if anything, they have an even wider structural diversity than the foreign chemicals that stimulate pulmonary C fibres and evoke pulmonary chemoreflexes. Histamine and serotonin are decarboxylated derivatives of amino acids, the prostaglandins are long chain unsaturated fatty acids and bradykinin is a large molecule, a nonapeptide split off from immunoglobulin A. Both histamine and serotonin act directly on bronchial smooth muscle, causing it to contract, but they have opposite effects on vascular smooth muscle, histamine being vasodilator and serotonin vasoconstrictor. The different prostaglandins have different direct effects on bronchial and vascular smooth muscle. Thus prostaglandins of the F series are bronchoconstrictor, whereas prostaglandins of the E series and prostaglandin I_2 are bronchodilator; generally speaking the effects of prostaglandins on vascular smooth muscle parallel their effects on bronchial smooth muscle, but they may vary somewhat with the vascular bed and the species. Bradykinin has a direct bronchoconstrictor action in some species, including guinea-pigs, but has no significant direct effect on bronchial smooth muscle in dogs or in man (*Waaler* 1961; *Garcia Leme* 1978). Both histamine and bradykinin are powerful vasodilators and increase capillary permeability in several species (*Garcia Leme* 1978; *Plaut* and *Lichtenstein* 1978).

4.1.2.1 Histamine

The respiratory effects of histamine have been investigated more frequently than have those of any other lung autocoid. Since asthmatic individuals are known to be highly sensitive to the bronchoconstrictor and irritant properties of histamine, studies of the chemical's respiratory effects were expected to provide some insight into the causative mechanisms of human asthma. Histamine has often been accepted as a specific stimulant of rapidly adapting (irritant) receptors in the conducting airways, and hence as the chemical of choice for studies of the reflex airway response for which irritant receptors were once thought to be primarily responsible (*Coleridge* and *Coleridge*, to be published). Nevertheless, results of cooling the vagus nerves in dogs (*Fishman* et al. 1973) and guinea-pigs (Koller and Ferrer 1973) suggest that the reflex respiratory effects of histamine also involve an afferent C fibre pathway. When histamine is injected intravenously in cats and dogs, however, it does not produce the prompt bradycardia typical of the pulmonary chemoreflex, an observation which by itself suggests that the stimulant effect of histamine on pulmonary C fibres is relatively minor; moreover, the latency of onset of respiratory effects in both cats (*Winning* and *Widdicombe* 1976) and dogs (*Coleridge* and *Coleridge*, unpublished observations) is longer than the pulmonary circulation time.

Afferent studies in general confirm that pulmonary C fibres are relatively insensitive to histamine (Fig. 7). *Armstrong* and *Luck* (1974) found in cats that right atrial injection of a large dose of histamine (50 µg/kg) stimulated 9 of 12 J receptors; however, although the response was sometimes prolonged, in other cases it consisted of no more than a single brief burst of firing. It was noteworthy that the response began 5.7 s, on average, after the end of the injection. While these results spoke in favour of a stimulant action of histamine on J receptors in cats, other observations in the same species were less positive. Thus *Paintal* found that histamine either failed to stimulate J receptors (*Paintal* 1969, 1970, 1973) or produced, after a long latency, effects so minor as to be insufficient to trigger the expected reflex response (*Paintal* 1974, 1977a, b). A similar lack of sensitivity to histamine on the part of pulmonary C fibres has been observed in dogs (*Kaufman* et al. 1980c). Injection of very large doses of

histamine (50–100 μg kg^{-1}) evoked no more than a few scattered impulses. The response, which occurred only after a long delay, coincided
with an increase in tracheal pressure and hence with the bronchoconstrictor effects of histamine; it was thought to be largely secondary to a decrease in lung compliance (see below).

Paintal (1974) has suggested that the response of J receptors (pulmonary C fibres) to histamine is due largely to mechanical changes, but
in his view such changes are secondary to the action of histamine in causing an increase in capillary permeability and exudation of fluid into the
alveolar interstitium. Whatever may be the mechanism by which histamine
activates pulmonary C fibres, there is no doubt that in dogs histamine has
much smaller effects on pulmonary C fibres than on bronchial C fibres.
Thus injection of 20 μg kg^{-1} histamine (a dose that has negligible effects

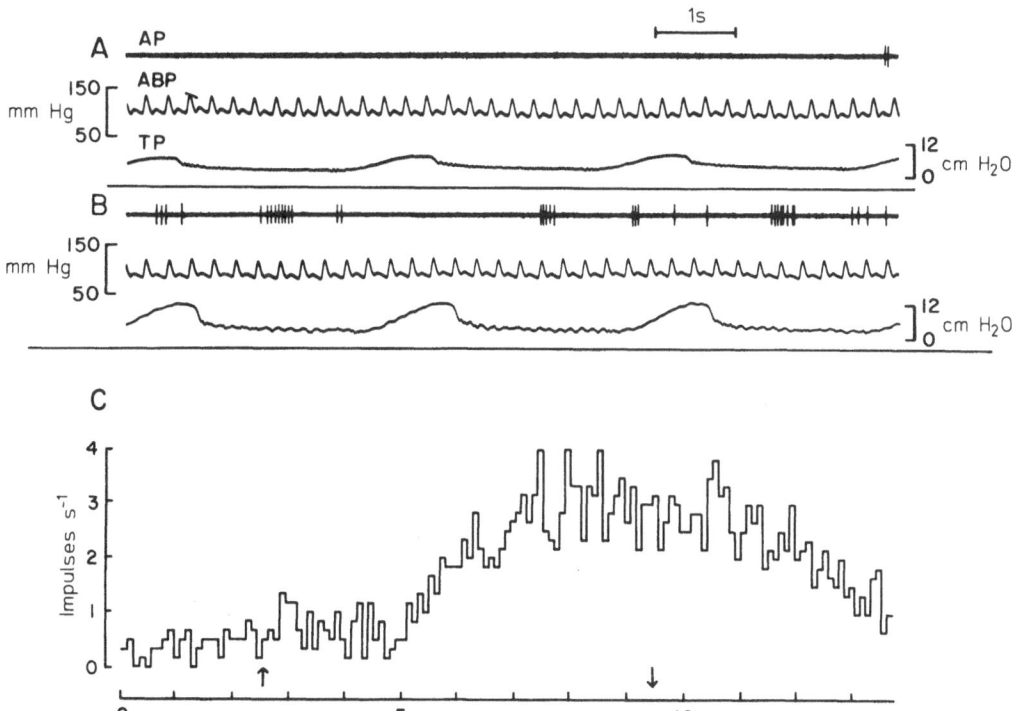

Fig. 9A–C. Stimulation of bronchial C fibres by administration of histamine aerosol
to the lower respiratory tract; experiments in two dogs with open chest and lungs
ventilated by a pump. **A, B** Impulse activity recorded from a bronchial C fibre (conduction velocity 1.3 m s^{-1}) arising from an ending in the right lung; **A** before and **B**
during inhalation of histamine aerosol (0.5%). *AP*, action potentials; *ABP*, arterial
blood pressure; *TP*, tracheal pressure. **C** Impulse frequency of a bronchial C fibre (conduction velocity, 1.6 m s^{-1}) arising from the right lung (different dog from that in
A and **B**); histamine aerosol administered between the *arrows*. Note irregular pattern
of discharge in **A** and **B**, and prolonged response in **C**. (*Coleridge and Coleridge* 1977b)

on pulmonary C fibres) into the right or left atrium (*Coleridge* et al. 1978a) or of 20 μg histamine into the bronchial artery (*Kaufman* et al. 1980c) vigorously stimulates bronchial C fibres, evoking the irregular bursts of firing that are so characteristic of the response of these afferent C fibres to both chemical and mechanical stimuli. Scattered bursts of impulses with no obvious relation to the ventilatory cycle are also evoked when histamine (0.1%–5.0%) is administered as an aerosol (Fig. 9) (*Coleridge* and *Coleridge* 1977b; *Coleridge* et al. 1978a).

4.1.2.2 Prostaglandins
Interest in the effects of prostaglandins on afferent endings in the lungs and airways stemmed from the observation that human subjects experienced symptoms of tracheal irritation and coughing while breathing aerosols of prostaglandins (*Herxheimer* and *Roetscher* 1971; *Kawakami* et al. 1973). Moreover the bronchodilator prostaglandins E_1 and E_2, delivered as aerosol (*Kawakami* et al. 1973; *Smith* and *Cuthbert* 1976) or as an intravenous infusion (*Smith* 1973), sometimes evoked a paradoxical bronchoconstriction, suggesting that a reflex originating in the airways was involved. Relatively large doses of prostaglandins $F_{2\alpha}$ ($2-12$ μg kg^{-1}) and E_2 ($5-10$ μg kg^{-1}) injected into the right atrium were found to stimulate both pulmonary and bronchial C fibres in dogs, PGE_2 being more potent in its effects than $PGF_{2\alpha}$ (Fig. 10) (*Coleridge* et al. 1976). The overall pattern of the afferent response to prostaglandins was very different from that evoked by capsaicin or phenyldiguanide, or even from that evoked by the other lung autocoids such as histamine or bradykinin, for both pulmonary and bronchial C fibres began to discharge only after a relatively long latency (PGE_2, mean 14.7 s, $PGF_{2\alpha}$, mean 24 s), and firing often continued for several minutes, suggesting that prostaglandins had a mechanism of action on the nerve endings quite different from that of phenyldiguanide and capsaicin. Enzymes that break down prostaglandins $F_{2\alpha}$ and E_2 are present in large amounts in the pulmonary vascular bed (*Piper* and *Vane* 1971). It was therefore not surprising that bronchial C fibres were stimulated more effectively when prostaglandins were injected into the left, rather than the right, atrium (*Coleridge* et al. 1978a; *Ginzel* et al. 1978), or directly into the bronchial artery (*Roberts* et al. 1981a).

Prostacyclin (PGI_2) is also a stimulant of pulmonary and bronchial C fibres in dogs (*Roberts* et al. 1981a). In rabbits, pulmonary C fibres appear to be highly sensitive to PGI_2: thus right atrial injection of PGI_2 (0.5 μg kg^{-1}) evokes effects comparable to those produced by very large doses (250 μg kg^{-1}) of phenyldiguanide, causing both pronounced stimulation of pulmonary C fibres and the immediate bradycardia and hypotension characteristic of the pulmonary chemoreflex (*Armstrong* and *Miller* 1981). In general, the effects of PGI_2 on pulmonary C fibres in rabbits differ

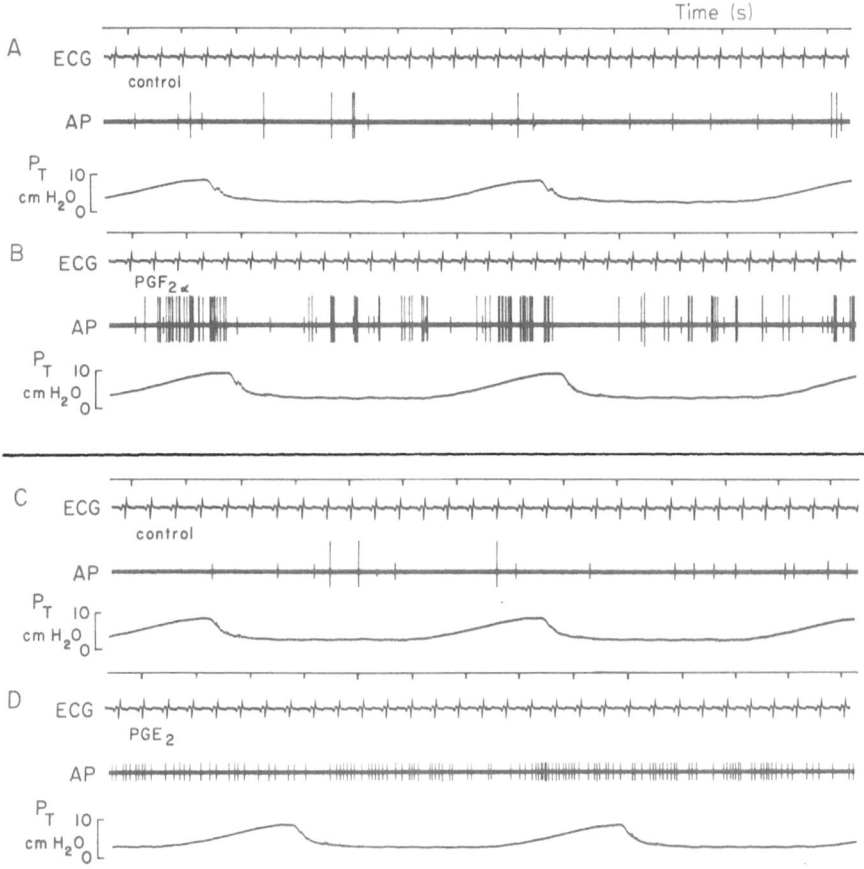

Fig. 10A–D. Response of a rapidly adapting (irritant) receptor (large spikes and a C fibre ending (small spikes) to prostaglandins. Dog, open-chest, lungs artificially ventilated. Both endings were located in the lower lobe of the left lung. **A** before and **B** 16 s after right atrial injection of prostaglandin $F_{2\alpha}$ (4 μg kg^{-1}). Interval of 6 min between **B** and **C**. **C** before and **D** 42 s after right atrial injection of PGE$_2$ (20 μg kg^{-1}). *From above downwards:* 1 s time trace; *ECG*, electrocardiogram; *AP*, action potentials recorded from a filament of the left vagus nerve; P_T, tracheal pressure. (*Coleridge* et al. 1976)

from those in dogs in having on average a much shorter latency and duration.

In dogs, administration of aerosols of 0.1% PGE$_2$ (*Coleridge* et al. 1978a), a concentration that produces marked symptoms of airway irritation in human subjects (*Smith* and *Cuthbert* 1976), or of 0.01% PGI$_2$ (*Roberts, Coleridge* and *Coleridge*, unpublished observations) stimulates airway C fibres, both bronchial C fibres within the lung and also similar endings in the lower trachea and extrapulmonary bronchi. Airway C fibres are also stimulated by cyclic ether analogues of the prostaglandin endoperoxide PGH$_2$, injected into the circulation or delivered as aerosol (*Ginzel* et al. 1978).

4.1.2.3 Bradykinin

Bradykinin clearly stimulates afferent nerve endings in the respiratory
tract, for in asthmatics and in some normal subjects it evokes cough, sensations of airway irritation and reflex bronchoconstriction when inhaled
as aerosol (*Lecomte* et al. 1962; *Simonsson* et al. 1973). Studies in dogs
have shown that the stimulant action of bradykinin in the lower respiratory tract is virtually confined to the endings of airway C fibres. Dose-
dependent effects on bronchial C fibres could be demonstrated after injection of bradykinin into the left atrium, injection of 1 μg kg^{-1} increasing activity 15-fold and often causing bursts of activity that lasted a minute
or more (Fig. 11) (*Kaufman* et al. 1980b). In contrast to its effects on
bronchial C fibres, however, bradykinin injected in similar doses into the
right atrium failed to stimulate pulmonary C fibres.

  Bradykinin, like the prostaglandins, is rapidly broken down in its
passage through the pulmonary circulation (*Levine* et al. 1973) but a sufficient amount remained in the bloodstream after right atrial injection to
have a prominent vasodilator action in these experiments (*Kaufman* et al.
1980b); hence it seems reasonable to conclude that active bradykinin was
still present in the pulmonary capillary blood, but that pulmonary C fibres
were relatively insensitive. Bronchial C fibres are stimulated most effective-

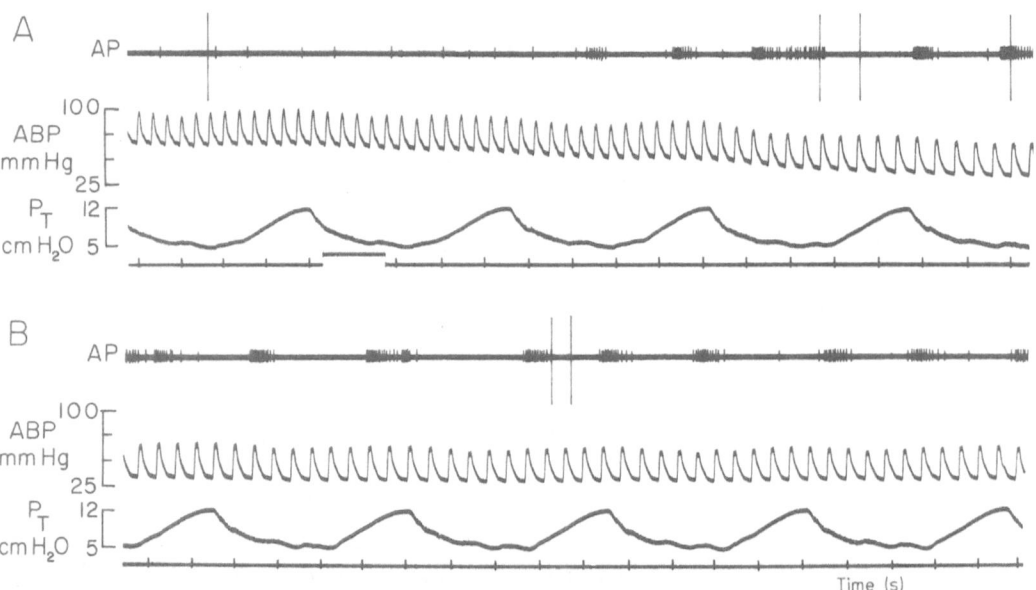

Fig. 11A, B. Effect of bradykinin on two airway endings in the left lung of a dog.
Impulses recorded from left vagal strand: one fibre (large spikes; conduction velocity,
21.9 m s^{-1}) arose from a rapidly adapting pulmonary stretch (irritant) receptor, the
other (small spikes, conduction velocity, 1.1 m s^{-1}) from a bronchial C fibre ending.
A bradykinin 1 μg kg^{-1} was injected into the left atrium at the signal (note decrease in
arterial blood pressure). Records **A** and **B** continuous. *AP*, action potentials; *ABP*,
arterial blood pressure; P_T, tracheal pressure. (*Kaufman* et al. 1980b)

ly by injecting bradykinin into the bronchial artery – 1.5 μg is sufficient to evoke a continuous high frequency discharge in some fibres and the characteristic bursts of activity in others (*Kaufman* et al. 1980; *Roberts* et al. 1981b), and smaller doses are often effective in evoking reflex effects. Injection of bradykinin into the bronchial artery in doses (0.02–1.5 μg) that stimulate bronchial C fibres and evoke conspicuous reflex effects causes no significant changes in firing of any other intrapulmonary vagal afferent (*Kaufman* et al. 1980b; *Roberts* et al. 1981b). Administration of bradykinin aerosol (0.1%) to the lower airways stimulates afferent C fibres with endings in the intrapulmonary and extrapulmonary bronchi; pulmonary C fibres are not affected (*Coleridge* et al. 1983).

One may conclude from these observations that differences in the response of pulmonary and bronchial C fibres to compounds of physiological interest are sufficiently pronounced to suggest that the two have somewhat different functions. Thus although both are strongly stimulated by exogenous irritants such as capsaicin, and although both are stimulated by the prostaglandins, the range of effective and endogenous chemicals is wider in the case of bronchial C fibres. Endogenous chemicals that stimulate bronchial C fibres are associated with inflammatory phenomena, and it is tempting to draw comparisons between bronchial C fibres and the cutaneous C fibres that respond to inflammatory mediators and cause sensations of itch and burning pain (*Lynn* 1977). Pulmonary C fibres, on the other hand, are more sensitive to mechanical changes in the lung parenchyma, and their physiological importance may lie in this particular aspect of their afferent properties (see below).

4.1.3 Response to CO_2

A physiological role for pulmonary C fibres as mixed-venous CO_2 sensors was first proposed by *Dickinson* and *Paintal* (1970), who briefly reported that J receptors in cats were often stimulated when a bubble of CO_2 was injected into the right ventricle, or when the lungs were briefly inflated with 20%–100% CO_2. *Russell* and *Trenchard* (1981) have since shown that pulmonary C fibres in rabbits are stimulated when sodium dithionite, a chemical that releases CO_2 from blood, is injected into the right atrium. Neither study has been published in full, however, hence the conclusion that a chemical action of CO_2 on nerve endings was the stimulating factor in each case cannot be accepted without reservation, and hence the functional importance of these observations remains difficult to assess. *Coleridge* et al. (1978b) found a very small, but statistically significant increase in the activity of pulmonary and bronchial C fibres (from 0.4 to 0.8 impulses s^{-1}) when PCO_2 in the vascularly isolated lung in dogs was increased from 19 to 30 mm Hg, but questioned the physiological significance of

the response because they obtained no further increase in firing upon rais-
ing lung PCO_2 to 46 mm Hg, and because firing did not revert to its original level when PCO_2 was reduced again. However, *Delpierre* et al. (1981)
have revived the concept of a physiological role for afferent C fibres as
CO_2 sensors, claiming that most 'bronchopulmonary C fibres' in cats are
CO_2 sensitive.

The experiments of *Delpierre* et al. were in cats with closed chest, the
lungs being ventilated by a pump; C fibre activity was examined when
end-tidal PCO_2, first reduced to 14 mm Hg by hyperventilation, was gradually increased to 70 mm Hg by administration of CO_2. Because the chest
was closed the intrapulmonary location of the endings was not established
by direct mechanical stimulation, and C fibres were selected entirely on
the basis of their response to injection of phenyldiguanide, being classified
as 'pulmonary' if stimulated within 5 s of right atrial injection and as
'bronchial' if stimulated within 5 s of the injection of phenyldiguanide
into the ascending aorta. However, such criteria would not exclude many
C fibres with endings sensitive to phenyldiguanide in the heart, great vessels
and other thoracic structures (*Coleridge* et al. 1973a; *Baker* et al. 1979)
nor would they exclude the chemoreceptors of the aortic bodies (*Paintal*
1967). (A few fibres were tested by administration of low O_2 mixtures

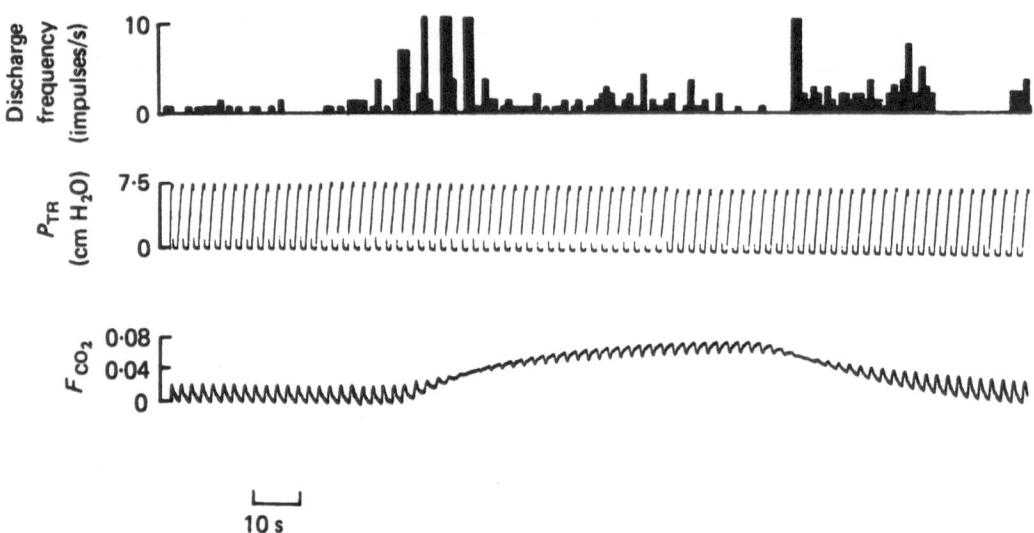

Fig. 12. Effect of hypercapnia on a pulmonary C fibre in a cat; chest intact, lungs
ventilated by a pump. Hypercapnia produced by adding CO_2 to the O_2 with which
the lungs were ventilated. Action potentials recorded in the nodose ganglion by means
of extracellular micro-electrodes; the afferent was identified as a pulmonary C fibre
because it was stimulated within 5 s of right atrial injection of phenyldiguanide. *From
above downwards:* discharge frequency of C fibre, counted by a rate-meter; P_{TR},
tracheal pressure; F_{CO_2}, tracheal concentration of CO_2. (*Delpierre* et al. 1981)

and since none was stimulated it was concluded that aortic chemoreceptors were absent from the larger sample.)

Delpierre et al. tested a large number of C fibres and 80% responded to CO_2 in some fashion. The discharge of the majority increased as end-tidal CO_2 increased from 14 to 28 mm Hg, but was unaffected thereafter, an overall response somewhat similar to that of pulmonary and bronchial C fibres in the non-perfused lung of dogs (*Coleridge* et al. 1978b, see above). The discharge of the remaining fibres continued to increase until end-tidal CO_2 reached a final steady level of 70 mm Hg, but it then adapted rapidly. These fibres responded to both 'on' and 'off' transients (Fig. 12), but although the investigators emphasize the potential significance of the transient responses, their physiological value in a pulmonary CO_2 sensor is not easy to assess. The chemoreceptor function of 'bronchopulmonary' C fibres would have been more convincingly demonstrated if even a small sample of fibres of confirmed pulmonary origin had been exposed to a stepwise increase in CO_2 over the physiological range. If a sensitivity to CO_2 is to be established as a property of lung C fibres, it must be on the basis of CO_2 response curves similar to those obtained for the arterial chemoreceptors (*Lahiri* et al. 1979). So far the results of afferent studies, though interesting, can be the basis for little more than speculation.

4.2 Response to Changes in Lung Volume

4.2.1 Response to Inflation

Pulmonary C fibres are stimulated by lung inflation; bronchial C fibres are also stimulated but their threshold is much higher (*Coleridge* and *Coleridge* 1977b; *Kaufman* et al. 1982a). The response of pulmonary C fibres to inflation was first demonstrated in dogs, with chest open and lungs artificially ventilated, by occluding the tracheal outlet so that the lungs were inflated progressively for two or three ventilatory cycles (Fig. 13). When the lungs are inflated in this fashion, a large majority of pulmonary C fibres in dogs begin to discharge at a lung volume between 1 and 2 V_T above FRC, and the remainder are recruited between 2 and 3 V_T above FRC (*Coleridge* et al. 1965; *Coleridge* and *Coleridge* 1977b). Similar inflation of the lungs by 2 or 3 V_T above FRC stimulates pulmonary C fibres in artificially ventilated cats (*Coleridge* et al. 1968; *Armstrong* and *Luck* 1974). We routinely use such inflation of the lungs as a simple, first step in the identification of pulmonary C fibres in vagal filaments in dogs and cats. *Paintal* (1969), who inflated the lungs of cats with unphysiologically large volumes (100–150 ml), reported stimulation of only an

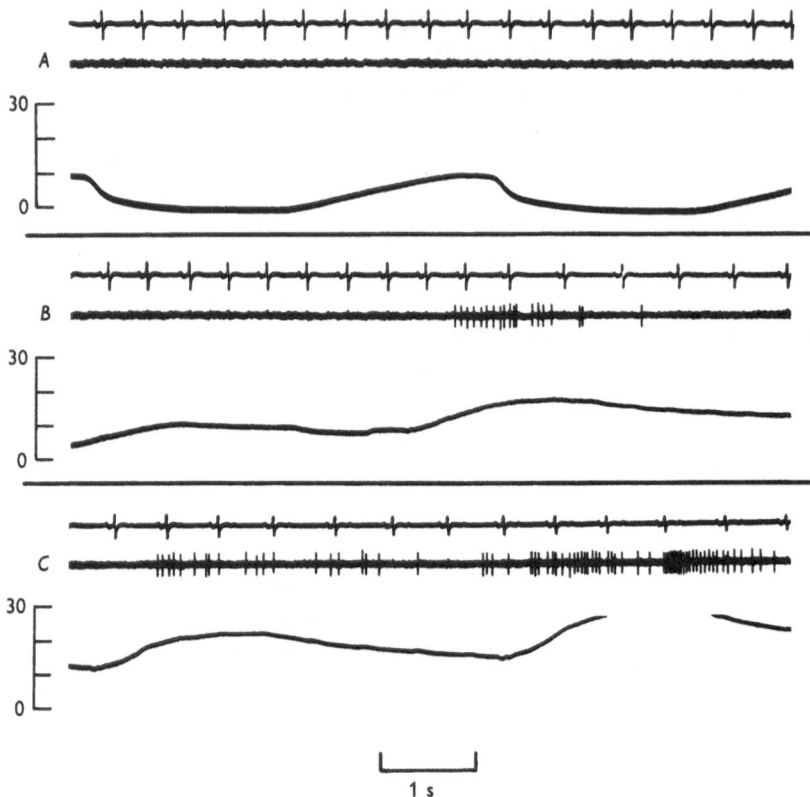

Fig. 13A–C. Stimulation of a pulmonary C fibre by lung inflation; dog, open chest, artificial ventilation. **A,B** and **C** are continuous. **A** Control; note that the fibre is silent. In **B** and **C**, the outlet tube from the tracheal cannula was clamped so that the lungs were progressively inflated by successive strokes of the ventilator. In **B**, the pulmonary C fibre was stimulated when tracheal pressure was about 13 cm H_2O and when lung volume was between FRC + 1 V_T and FRC + 2 V_T, i.e. when lung volume was within the physiological range. Note cardiac slowing during inflation. *From above downwards:* electrocardiogram; vagal action potentials, tracheal pressure. (*Coleridge* et al. 1965)

occasional J receptor and concluded that this group of afferents was in-sensitive to inflation. Leaving *Paintal's* results aside, the evidence on this aspect of pulmonary C fibre function is quite definite, and there is no doubt that the endings are stimulated by lung inflations that are large but well within the physiological range, and that the reflex depression of heart rate evoked by large lung inflations arises mainly from activation of pul-monary C fibres (*Cassidy* et al. 1979; *Kaufman* et al. 1982a; *Coleridge* and *Coleridge,* to be published).

The typical response of pulmonary C fibres to successive tidal volumes consists of a scattered train of impulses with each sequential inflation, activity often appearing to 'adapt' towards the end of each inflation cycle. This apparent adaptation may be in part due to the method used to inflate

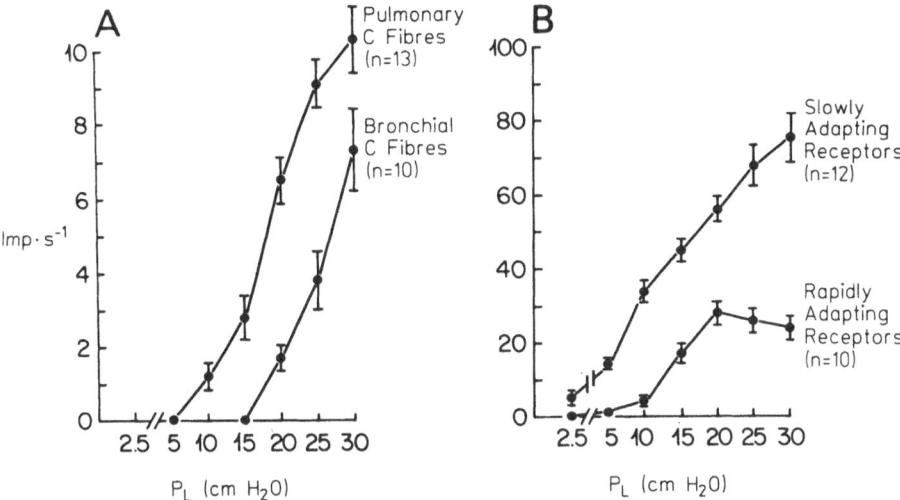

Fig. 14A, B. Stimulation of pulmonary vagal afferents by lung distension in dogs; response of A non-myelinated and B myelinated fibres. The chest was open and the right and left lungs were ventilated separately. The left pulmonary artery was ligated so that the right lung received the entire right ventricular output. Afferent impulses originating in the left lung were recorded from the left cervical vagus nerve. Stimulus response curves were obtained as the left lung was slowly inflated to a pressure of 30 cm H_2O at a rate of 1.5–2.0 cm H_2O s^{-1}. Results are means ± SE; P_L, left airway pressure. (*Kaufman* et al. 1982a)

the lung (i.e. a volume stimulus) and to the likelihood that pulmonary C fibres, in their role as 'high threshold inflation receptors', respond like slowly and rapidly adapting stretch receptors to the tension in the lung tissue in which they are embedded, so that their discharge is related to transpulmonary pressure rather than to lung volume per se. *Kaufman* et al. (1982a), in experiments in dogs, have used a more carefully controlled method to examine the threshold and sensitivity of lower airway afferents to lung inflation. They increased transpulmonary pressure in a non-perfused lung from a baseline of 2.5 cm H_2O to a maximum of 40 cm H_2O at a rate of approximately 2 cm H_2O s^{-1} (a rate of inflation which probably corresponds to that during quiet breathing). The results of *Kaufman* et al. indicate that the threshold of pulmonary C fibres to a slow ramp of inflation is similar to that of rapidly adapting (irritant) receptors (Fig. 14). Whether pulmonary C fibres are also sensitive to the rate of inflation is unknown at present.

Bronchial C fibres are less responsive than pulmonary C fibres to lung inflation. The difference is well illustrated by the results of experiments in which *Coleridge* and *Coleridge* (1977b) compared effects on eight fibres of each type when the lungs of open chest dogs were inflated by successive strokes of the ventilator. Pulmonary C fibres responded with an average of

18 impulses at 2 V_T and 34 impulses at 3 V_T. In these experiments V_T was in the normal range for dogs (15 ml kg^{-1}), and FRC was the volume of air in the lungs at a transpulmonary pressure of 3–4 cm H_2O. By contrast, only three of eight bronchial C fibres were stimulated; they fired with an average of two impulses at 2 V_T and seven impulses at 3 V_T. Kaufman et al. (1982a), using their more controlled method of inflation, found that the threshold transpulmonary pressure required to stimulate bronchial C fibres was 10 cm H_2O higher than that needed to stimulate pulmonary C fibres, although the slopes of the two stimulus–response curves were parallel (Fig. 14).

4.2.2 Response to Deflation

There is little evidence that deflation stimulates lung C fibres. Neither bronchial C fibres in dogs (Coleridge and Coleridge 1977b) nor pulmonary C fibres in dogs and cats (Coleridge et al. 1965; Armstrong and Luck 1974) are stimulated when the lungs are allowed to collapse in the open chest, or when air is sucked from the trachea during spontaneous breathing. Armstrong and Luck, who identified J receptors (pulmonary C fibres) in cats by the criteria recommended by Paintal, found no evidence that deflation either stimulated the receptors or sensitized them to chemical stimulation by phenyldiguanide, as Paintal has described in his initial study (Paintal 1955); all but one of the receptors was stimulated by lung inflation, however. Paintal has since concluded that lung deflation is at best a weak stimulus to J receptors and that indeed many are unaffected by deflation (Paintal 1969).

4.3 Response to Pulmonary Vascular Changes

When Paintal re-evaluated the afferent properties of what he had originally called 'deflation receptors' he postulated that the newly entitled 'juxtapulmonary (type J) receptors' were interstitial stretch receptors whose natural stimulus was an increase in alveolar interstitial pressure or volume, caused by an increase in pulmonary capillary pressure, and whose physiological function was to signal the pulmonary vascular changes of exercise (Paintal 1969, 1970). The experimental basis for Paintal's new hypothesis rested mainly on his observation that J receptor activity was increased by obstruction of left ventricular outflow and by insufflation of chlorine and intravenous injection of alloxan, two chemicals that produce marked pulmonary congestion and oedema. However, the part played by pulmonary vascular changes in the stimulation of J receptors by alloxan and chlorine is not easy to estimate, because these chemicals are irritant and may stimulate nerve endings, not only in the lung itself but also at sites in the systemic circulation, by a purely chemical action (*Coleridge* and *Coleridge* 1977a).

The possibility of a reflex role for pulmonary C fibres in exercise is more strongly supported by the recent observations of *Anand* and *Paintal* (1980) in cats that six of ten J receptors in the right lung increased their discharge on average from 0.06 to 0.9 impulses s^{-1} when the left pulmonary artery was occluded to increase blood flow through the right lung.

Indirect evidence that pulmonary C fibres may be sensitive to the volume of blood in the lungs, or to the rate of blood flow through the lungs, is also provided by the observation that in both dogs and cats these C fibres fire at significantly higher rates during spontaneous breathing than they do after the chest has been opened and the lungs ventilated artificially at a similar FRC, V_T and frequency (*Coleridge* and *Coleridge* 1977a). Both cardiac output and central blood volume decrease when the chest is opened and the lungs are ventilated with positive pressure, hence sensitivity to pulmonary circulatory changes may account for the difference in pulmonary C fibre discharge in the two conditions. The activity of bronchial C fibres, however, was not significantly different in the closed and open chest (*Coleridge* and *Coleridge* 1977b).

An increase in blood flow through the lungs is an undeniable accompaniment of exercise; pulmonary congestion and oedema, on the other hand, are pathophysiological changes that occur in exercise only in exceptional circumstances. Acceptance of the hypothesis that pulmonary C fibre activity increases significantly in exercise would be greatly strengthened if these C fibres could be shown to be sensitive to controlled changes in pulmonary blood flow. Such evidence has yet to be obtained.

Both pulmonary and bronchial C fibres are undoubtedly sensitive to pulmonary congestion and are stimulated when left atrial pressure, and hence pressure in the pulmonary vascular bed, is increased progressively by distension of a small balloon floating in the mitral orifice (*Coleridge* and *Coleridge* 1975, 1977a). When pulmonary congestion was induced in this way, individual pulmonary and bronchial C fibres were found to be highly responsive, but on average pulmonary C fibres were more sensitive (Fig. 15). Thus one-third of pulmonary C fibres were stimulated at left atrial pressures of no more than 5 mm Hg above the control value, and 80% were stimulated when left atrial pressure had increased to 20 mm Hg above control; by contrast only a third of bronchial C fibres were stimulated by an increase in left atrial pressure of 20 mm Hg, and a third were unaffected even when left atrial pressure had increased to about 30 mm Hg (*Coleridge* and *Coleridge* 1975). The most sensitive fibres in both categories began to fire with a. ventilatory rhythm as left atrial pressure increased (Fig. 15); others fired irregularly throughout. Certainly the stimulation of

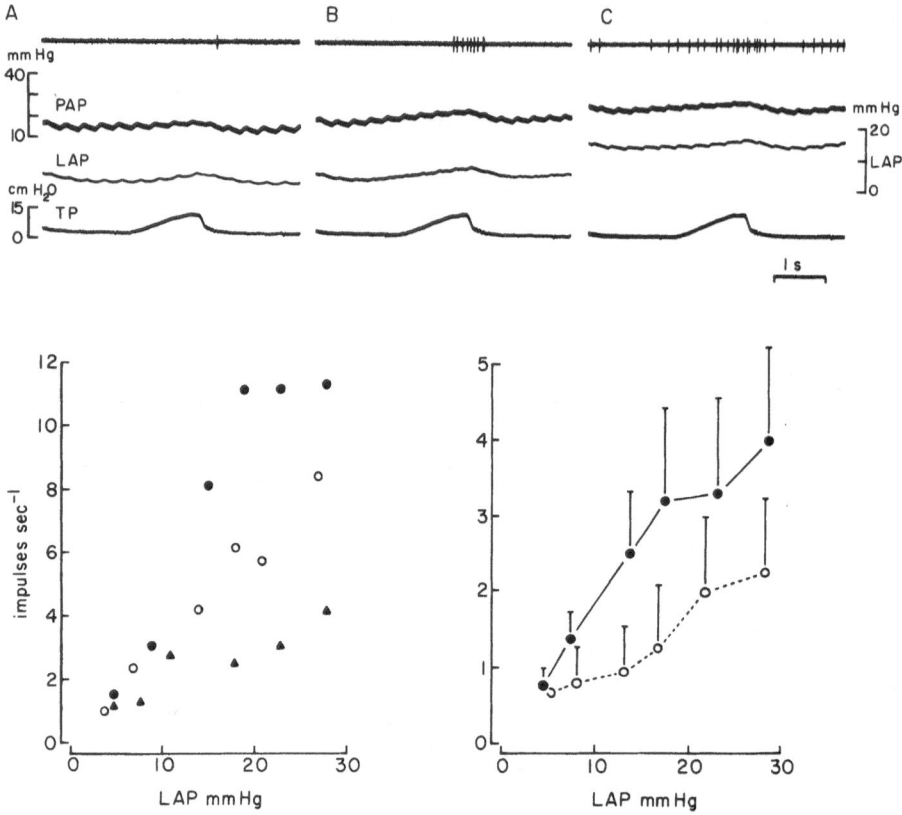

Fig. 15. Stimulation of lung C fibres by pulmonary congestion in dogs with open chest and lungs ventilated by a pump. C fibre impulses recorded from strands of the cervical vagus nerve. Left atrial and pulmonary vascular pressures were increased in steps by distending a small balloon placed in the left atrioventricular orifice. *PAP*, pulmonary arterial pressure; *LAP*, left atrial pressure; *TP*, tracheal pressure. *Above,* progressive stimulation of a pulmonary C fibre by lung congestion. *Below: on left,* activity of three pulmonary C fibres as function of mean left atrial pressure (each point represents average impulse frequency counted over 20 s); *on right,* activity of ten pulmonary C fibres (●) and eight bronchial C fibres (○) as function of left atrial pressure; results are means ± SE. (Based on data reported by *Coleridge* and *Coleridge* 1977a)

pulmonary C fibres in these experiments seemed sufficient to account for the pulmonary chemoreflex-like effects (apnoea, rapid shallow breathing, bradycardia and hypotension) observed at the onset of acute severe experimental pulmonary congestion in cats (*Churchill* and *Cope* 1929) and dogs (*Aviado* et al. 1951; *Downing* 1957).

4.4 Pulmonary Embolism and Inflammation

4.4.1 Pulmonary Embolism

The association of a vagally mediated tachypnoea with pulmonary embolism, and the persistence of this tachypnoea when the vagus nerves were cooled to low temperatures (*Whitteridge* 1950), prompted *Paintal* (1955) to examine the effects on 'deflation receptors' of injecting potato starch granules into the pulmonary circulation. This procedure induced multiple small action potentials in vagal filaments that contained the fibres of 'deflation receptors' stimulated by phenyldiguanide, and *Paintal* concluded that the tachypnoea of pulmonary embolism could probably be ascribed to stimulation of 'deflation receptors'. The effectiveness of pulmonary embolism (potato starch or small glass or plastic beads) as a stimulus to pulmonary C fibres in cats has since been confirmed by *Paintal* et al. (1973) and *Armstrong* et al. (1976). Similar effects on pulmonary C fibres have been described in rabbits (*Armstrong* and *Miller* 1980).

The nature of the stimulus to the nerve endings in pulmonary embolism is probably complex, and embolism is unlikely to be effective simply because it causes local mechanical distortion of small pulmonary vessels (*Paintal* et al. 1973; *Armstrong* et al. 1976); release of chemical mediators, particularly serotonin, by breakdown of platelets is a likely mechanism of the afferent stimulation (*Dawes* and *Comroe* 1954; *Widdicombe* 1964). Experimental support for this hypothesis has been obtained in rabbits, in which effects of embolism on pulmonary C fibres and myelinated lung afferents were significantly attenuated after an experimentally induced depletion of platelets (*Armstrong* et al. 1979; *Armstrong* and *Miller* 1980). In these experiments induction of platelet breakdown by injection of anti-serum resulted in stimulation of afferent fibres, thus providing persuasive if indirect evidence that the release of serotonin from platelets plays a major part in the afferent effects of embolism.

Since bronchial C fibres are stimulated by serotonin, they are likely to contribute to the increase in C fibre input in embolism. Indeed, in dogs and possibly also in man, bronchial C fibres may play a more important role than pulmonary C fibres, which are relatively insensitive to serotonin in these two species (*Dawes* and *Comroe* 1954; *Kaufman* et al. 1980c). Although serotonin is probably the major chemical factor responsible for the respiratory changes induced by pulmonary embolism, other autocoids must also be involved, since anti-serotonin agents do not block the responses completely (*Halmagyi* and *Colebatch* 1961). Release of other lung autocoids is likely to contribute to the afferent effects of pulmonary embolism.

4.4.2 Inflammation

There can be little doubt that afferent C fibres from the lungs and airways are stimulated by inflammation of the lung. As in the case of embolism, the nature of the stimulus to the nerve endings is likely to be complex, and speculation as to the precise mode of stimulation seems unlikely to be fruitful at present. In experiments in cats, *Frankstein* and *Sergeeva* (1966) demonstrated a considerable increase in afferent vagal C fibre activity during acute pulmonary inflammation induced by lobar injection of 40% glucose solution and hot water. Using the collision method of *Douglas* and *Ritchie* (1962) to assess the total input of C fibres in branches from the lower airways, they were unable to detect a respiratory variation in the heightened discharge. However, we have occasionally observed conspicuous trains of impulses during inflation in pulmonary C fibres supplying patches of lung of unhealthy inflamed appearance; these were no more than chance observations made in the source of other experiments, in which the presence of patchy pneumonitis was an unwelcome complication rather than a matter of experimental design. Our observations were merely incidental therefore; and since it was the phasic nature of the C fibre discharge that brought it to our attention, many other C fibres may have fired irregularly and been overlooked.

5 Reflexes Triggered by Lower Respiratory Tract C Fibres

The history of the reflexes evoked by stimulation of afferent C fibres in the lower respiratory tract begins at the end of the nineteenth and the beginning of the twentieth centuries, when investigators had at last abandoned attempts to explain vagal reflex changes in breathing solely in terms of changes in activity in a single class of stretch or inflation receptor that inhibited inspiration. From this time the existence of other pulmonary afferents with different reflex functions was increasingly recognized, largely because stimuli applied to the lower respiratory tract were found to cause reflex changes in breathing of vagal origin after the classical Hering-Breuer inflation and deflation reflexes had been abolished by cooling or compressing the vagus nerves in the neck. The use of chemical stimuli played a large part in drawing attention to the possible reflex contribution of fine afferent fibres to lower airway reflexes (*Brodie* 1900; *Brodie* and *Russell* 1900; *Dawes* and *Comroe* 1954), and chemicals continue to play an important role in defining the properties of both pulmonary and bronchial C fibres. In addition, analysis of the reflex response to inflation of the lung during partial vagal block provided early evidence for a small-fibre afferent input with marked effects on breathing (*Head* 1889; *Hammouda* and *Wilson* 1935a, b, 1939).

5.1 Introduction to Reflexes Evoked by Chemicals

5.1.1 Nomenclature

Before the work of *Brodie* (1900), investigators believed that the respiratory tract below the vocal chords was insensitive to chemical irritants. *Brodie* provided the first description of what were later to be called the pulmonary chemoreflexes (*Dawes* and *Comroe* 1954). He described a reflex 'triad' of apnoea, bradycardia and systemic hypotension evoked in cats with 'great suddenness' when serum or egg-white was injected into a jugular vein. *Brodie* and *Russell* (1900) observed that a similar reflex 'triad' was evoked in dogs when high concentrations of irritant gases were delivered to the lower airways, and they suggested that the same afferent pathway was involved in both cases. However, we would now consider that the reflex response to inhalation of airway irritants described by *Brodie* and *Russell* falls outside the definition of pulmonary chemoreflexes given by *Dawes* and *Comroe* (1954), in that it did not necessarily arise from nerve endings immediately accessible from the pulmonary circulation. Instead the effects described by *Brodie* and *Russell* seem to provide an example of the more broadly defined airway defence reflexes (*Widdicombe* 1974b, 1977a, 1981; *Coleridge* and *Coleridge* 1981), which are evoked by administration of particulate and chemical irritants to the airways and which include apnoea, cough, gasps, rapid shallow breathing, bronchoconstriction and increased airway secretion. The airway defence reflexes are the outcome of changes in activity in both myelinated and non-myelinated afferent vagal fibres, possibly, although not necessarily, including the pulmonary C fibres responsible for the pulmonary chemoreflexes. The pulmonary chemoreflexes are defined as the constellation of reflex effects (Fig. 16) evoked by the action of certain chemicals in the pulmonary vascular bed, and the lower airway defence reflexes as the constellation of effects evoked when irritants gain access to the lower airways. (Strictly speaking, of course, the pulmonary chemoreflexes could be included in the more broadly defined lower airway defence response, because pulmonary C fibres may sometimes contribute to the total afferent vagal input evoked by irritants entering the lower airways. However, in the present account we follow the usual custom of regarding the pulmonary chemoreflexes as a separate entity.)

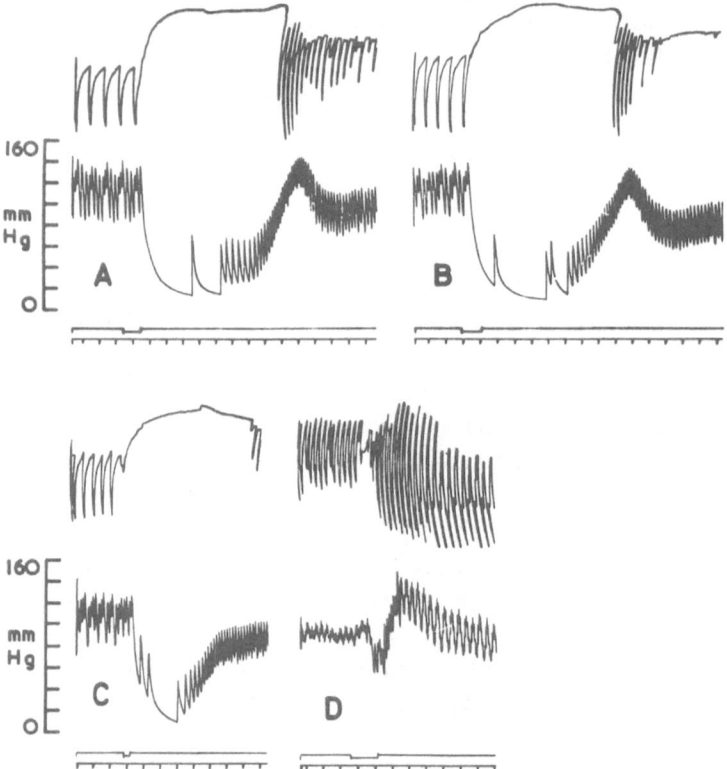

Fig. 16A—D. Pulmonary depressor chemoreflex elicited by injecting capsaicin (20 μg kg^{-1}) into bloodstream of spontaneously breathing anaesthetized dog; vagus nerves intact. Capsaicin was injected at various sites, as follows: **A** into a femoral vein; **B** through a catheter whose tip was in the origin of main pulmonary artery; **C** through a catheter into left pulmonary artery, just beyond origin of first lobar branch; **D** into left atrium through a needle-tipped cannula inserted via the trachea. In each record: *above*, breathing recorded with a pneumograph (inspiration downwards); *below*, arterial blood pressure, time trace, 5 s. Note marked similarity of effects produced by the three injections made upstream to the lungs (**A**, **B** and **C**), and the quite different pattern of response produced by left atrial injection (**D**). (*Coleridge* et al. 1964b)

5.1.2 Chemicals That Evoke Pulmonary Chemoreflexes

The pulmonary chemoreflex response has been investigated in rabbits, cats, dogs, rats and man, and the reader is referred to the review by *Dawes* and *Comroe* (1954) for an extensive bibliography of the subject. The wide variety of effective chemicals listed by *Dawes* and *Comroe* includes amid-ines (of which the most frequently used is phenyldiguanide), isothioureas, antihistamines, serotonin, nicotine, ammonia, ATP and lobeline; to these must now be added capsaicin (*Toh* et al. 1955; *Porszasz* et al. 1957; *Coleridge* et al. 1964b) and certain opiate polypeptide analogues (*Sapru* et al. 1981). The action of these chemicals appears to depend on specific pharmacological properties of the nerve endings, rather than, as was once believed, on a pronounced sensitivity of afferent C fibres to chemicals in general (*Paintal* 1964). For example, phenyldiguanide and serotonin have certain structural features in common, and in the occasional cat in which phenyldiguanide fails to evoke the pulmonary chemoreflex, serotonin is equally ineffective. Moreover, certain other amidines and bufotene, which have structural features in common with phenyldiguanide and serotonin, block the reflex effects of both these compounds (*Fastier* et al. 1959). Species differences in susceptibility to the chemicals that evoke the pulmonary chemoreflexes are not unusual, as *Brodie* (1900), who found serum and egg-white effective in cats but not in dogs or rabbits, was the first to observe. Thus phenyldiguanide evokes the pulmonary chemoreflex in cats (*Dawes* et al. 1951; *Paintal* 1955, 1957; *Fastier* et al. 1959), rabbits (*Dawes* et al. 1951; *Karczewski* and *Widdicombe* 1969c) and rats (*Sapru* et al. 1981), but not in dogs (*Dawes* et al. 1952; *Coleridge* and *Coleridge* 1977b) or man (*Jain* et al. 1972). Capsaicin is most commonly used to evoke the pulmonary chemoreflex in dogs (Fig. 16) (*Coleridge* et al. 1964b; *Brender* and *Webb-Peploe* 1969); it is also effective in cats (*Toh* et al. 1955) and rats (*Sapru* et al. 1981). Lobeline evokes the pulmonary chemoreflex in man (*Jain* et al. 1972).

With few exceptions, the chemicals used to evoke pulmonary chemoreflexes have, in the doses commonly employed, no physiological actions except those evoked reflexly by stimulation of afferent nerve endings. Although the various components of the pulmonary chemoreflex response are by definition those attributable to stimulation of afferent vagal endings immediately accessible from the pulmonary vascular bed (i.e. pulmonary C fibres), additional nerve endings, located further downstream and recruited when the chemicals reach the systemic circuit, may eventually complicate the reflex picture (Fig. 5A). Even so, interpretation of the reflexes evoked by chemicals is often simpler than interpretation reflexes evoked by changes in lung volume, which may have marked effects on the circulation and cause changes in the blood gases. Injection of chemicals such as phenyldiguanide and capsaicin remains a favourite method for studying the reflex properties of pulmonary C fibres, even though the response evoked by sudden injection of a bolus of foreign chemical into the right heart or pulmonary artery may with some justification be regarded as an experimental curiosity which has no counterpart outside the laboratory. Few chemicals of intrinsic physiological interest are known to evoke the pulmonary chemoreflex triad. Serotonin, which is one of the examples, evokes the chemoreflex triad in cats, but not in dogs or in man (*Dawes* and *Comroe* 1954).

5.1.3   Chemicals That Evoke Airway Defence Reflexes

The chemicals that evoke airway defence reflexes are a different matter, for they include not only foreign chemicals such as ammonia, bromine and sulphur dioxide, but also endogenous chemicals such as histamine, bradykinin and the prostaglandins that are known to be formed and released in the walls of the lower airways and to have a variety of physiological actions. Both histamine and bradykinin produce disturbances of breathing and contraction of airway smooth muscle in dogs when injected directly into the bronchial circulation (*DeKock* et al. 1966; *Roberts* et al. 1981b; *Coleridge* et al. 1983). This extension of the classical chemoreflex approach has been used in dogs to investigate the reflex properties of afferent C fibres in the conducting airways; bradykinin has been the lung autocoid of choice, because, unlike histamine, it is without direct bronchoconstrictor effects in dogs, and in small doses stimulates only bronchial C fibres. (Bradykinin has a powerful direct bronchoconstrictor action in guinea-pigs, which makes it less useful for investigation of airway reflexes in this species.)

5.2   Introduction to Reflexes Evoked by Lung Inflation

The application of chemical stimuli to the lower respiratory tract by *Brodie* and *Russell* in 1900 can be regarded as an early milestone in the study of reflexes evoked by the C fibre pathway. An even earlier discovery, one that in our view has been subject to considerable misinterpretation, has implications for the functions of lung C fibres that can only now be fully appreciated. This was the 'paradoxical effect' of lung inflation described by *Head* (1889).

5.2.1   Head's Paradoxical Reflex in Rabbits

This 'paradoxical' response, which consisted of a reversal of the usual Hering-Breuer inhibitory response to lung inflation, was evoked in rabbits whose vagus nerves were rewarming after being packed in ice (Fig. 17) (*Head* 1889). Head stressed that the vigour of diaphragmatic contraction during the paradoxical inspiration in rabbits was roughly the same as that of a normal inspiration, and he warned against the use of large or abrupt inflations, which were known to evoke sudden gasps. Nevertheless, what is now known as "Head's paradoxical reflex" is almost invariably equated with the 'gasp reflex', a brief powerful contraction of inspiratory muscles evoked by a large, abrupt inflation (*Widdicombe* 1954b; *Reynolds* 1962), and Head's original description of the paradoxical effect in rabbits is quite disregarded.

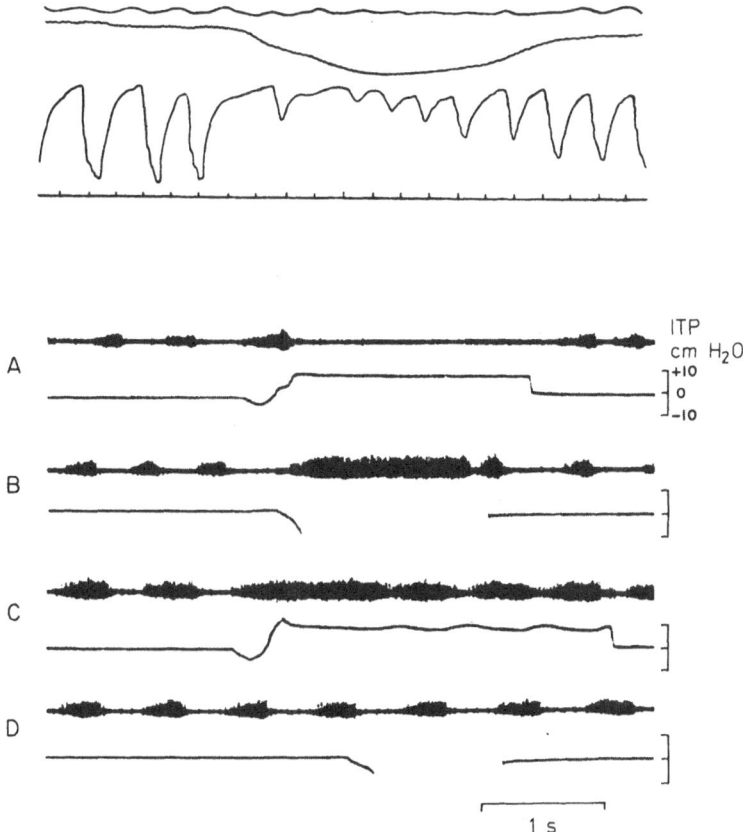

Fig. 17. Head's paradoxical reflex. *Above:* Head's original record depicting diaphragmatic contraction evoked by moderate lung inflation in a rabbit whose vagus nerves were rewarming after having been packed in ice (lungs inflated during fourth respiratory cycle). When the vagi were at body temperature, inflation evoked diaphragmatic relaxation. *From above downwards:* movements of chest wall; arterial blood pressure, contraction of diaphragmatic slip (contraction upwards). *Below:* Effect of inflating *(A, C)* and deflating *(B, D)* the lungs on the diaphragmatic electromyogram *(upper trace)* and intratracheal pressure *(ITP; lower trace)* in a rabbit. *A, B,* vagi at body temperature: *A,* inflation causes inhibition of diaphragmatic activity; *B,* deflation increases activity. *C, D,* vagi cooled to 5°C: *C,* inflation now increases diaphragmatic activity (Head's paradoxical reflex); *D,* deflation has no effect. *(Above, Head* 1889; *below, Widdicombe* 1967)

Subsequent investigators have confirmed Head's observations and have shown that the paradoxical effect can be evoked in rabbits when the vagus nerves are cooled to 5°C (Fig. 17) (*Widdicombe* 1967) and that it is present at a temperature as low as 3°C (*Whitteridge* and *Bulbring* 1944). At these temperatures, the normal inhibitory response to moderate lung inflation is replaced by a tonic inspiratory effort on which rapid shallow breathing movements are superimposed, and the classical Hering-Breuer deflation reflex is abolished (Fig. 17).

The identity of the vagal afferents responsible for Head's paradoxical reflex in rabbits has been the subject of a good deal of controversy over the years. *Paintal* (1966) labelled the response a 'physiological artefact' and suggested that it arose not from a separate group of small afferent fibres that were able to conduct at low temperatures, but merely from a differential block of the high frequency component of slowly adapting stretch receptor discharge and the consequent depression and phase reversal of their input. *Widdicombe* (1967) refuted this suggestion, and in one of the few studies in which the blocking temperature for afferent fibres has been recorded in the same experimental preparation as the blocking temperature for a reflex, showed in rabbits that impulses from pulmonary stretch receptors were totally blocked at the temperature at which the Hering-Breuer inhibitory reflex was abolished (8°–12°C); by contrast Head's paradoxical reflex could still be obtained at temperatures several degrees lower (Fig. 17).

There can be little doubt from what we now know of the cooling temperatures required to block activity in myelinated fibres that Head's paradoxical reflex is initiated by activation of non-myelinated fibres. In rabbits a similar inspiratory effort with rapid shallow breaths at increased FRC is characteristic of the respiratory component of the pulmonary chemoreflex and survives selective blockade of myelinated vagal fibres (*Dawes* et al. 1951; *Karczewski* and *Widdicombe* 1969c; *Guz* and *Trenchard* 1971). A similar response is evoked in rabbits when the central end of the vagus nerve is stimulated, below a point at which conduction in myelinated fibres is blocked selectively by anodal polarization (*Guz* and *Trenchard* 1971). Hence this pattern of respiratory response appears to be typical of stimulation of afferent C fibres in the rabbit, and discussion of its functional significance is best confined to the rabbit. One must conclude from the above observations that the afferent limb of Head's paradoxical reflex in rabbits is carried in C fibres, and hence that pulmonary C fibres in rabbits are activated by lung inflation, as they are known to be in cats (*Coleridge* et al. 1968; *Armstrong* and *Luck* 1974) and dogs (*Coleridge* et al. 1965; *Coleridge* and *Coleridge* 1977b; *Kaufman* et al. 1982a).

5.2.2 Effects of Inflation in Other Species

Reflex effects on breathing that differed from the classical inhibition of inspiration were observed by *Hammouda* and *Wilson* (1935a, b, 1939) in dogs when lung inflation was combined with partial vagal blockade. Thus, a moderate lung inflation that evoked prolongation of the expiratory pause under control conditions caused rapid breathing movements when the vagus nerves were compressed or cooled; effects were present at vagal temperatures between 8°C and a few degrees above freezing point, and were said to be maximal at a vagal temperature of 5°C. Although a tonic increase in inspiratory activity was not a feature of the response in dogs, as it was in rabbits, *Hammouda* and *Wilson* (1935b) interpreted the effects as being essentially the same reflex phenomenon as Head's paradoxical reflex and as revealing the existence of a group of small afferent fibres from the lungs, which they called 'respiratory−accelerator fibres'.

In a study of the reflex effects of lung inflation in cats, *Widdicombe* (1954b) found that a sudden large inflation that evoked the 'gasp reflex' under control conditions evoked instead a brief apnoea, termed a 'small fibre inhibitory reflex', when the vagus nerves were cooled to 7°−8°C. With benefit of hindsight the apnoea can be interpreted as the result of an abrupt surge of activity in pulmonary C fibres, which accords with the observation that apnoea is the typical respiratory response in cats when pulmonary C fibres are abruptly and strongly stimulated by bolus injections of chemicals.

Taking these early results at their face value, without reference to the type of inflation used in a given case, one might conclude that inflation of the lung during vagal cooling evokes quite different effects in the three species, causing a tonic inspiratory drive and accelerated breathing in rabbits, accelerated breathing in dogs and inhibition of breathing in cats (all effects that now appear attributable to stimulation of C fibres). Small wonder, therefore, that these early experiments on the vagal control of breathing have left behind an impression of confusion and have been interpreted by some as evidence of the unreliability of differential cooling as a method for examining respiratory reflexes, and by others as evidence of species differences in vagal reflex responses that are so marked as to make it impossible to derive any general principles from such observations. As we shall see, however, stimulation of pulmonary C fibres by inflation of the lung has provided important clues to the functional significance of these afferents. Moreover, the differential nerve blocking methods already in use in the time of Hering are still extremely useful, although now employed with a number of refinements and with a clearer understanding of the principles involved. The method of combining a physiological stimulus (which may change the activity of several types of afferent), with selective blockade of myelinated fibres, as exemplified in Head's study, still occupies a central position in the investigation of reflexes from the lower respiratory tract.

5.3   Methods for Selective Vagal Block

Methods for producing differential nerve block have been applied to the study of vagal respiratory reflexes perhaps more frequently than to any other neural control system, and no account of the reflexes evoked by stimulation of lower respiratory tract C fibres would be complete without some brief consideration of their usefulness and limitations. Three methods, cooling, anodal polarization and compression, are used to block conduction selectively in the myelinated fibres of a nerve trunk while sparing that in non-myelinated fibres; application of local anaesthetic is used to block conduction in non-myelinated fibres while sparing that in myelinated ones. The differential conduction block achieved by cooling and anodal polarization appears to depend on selective blockade of saltatory conduction (*Paintal* 1965; *Franz* and *Iggo* 1968; *Casey* and *Blick* 1969; *Whitwam* and *Kidd* 1975). Compression of nerves as a method of differential block is now rarely used except in studies of limb sensation in man (*Torebjork* and *Hallin* 1974); it appears to block myelinated fibres selectively, but whether as a function of fibre diameter or of saltatory conduction is not known. Tension on nerve trunks has effects similar to compression, and any tension accidentally exerted during cooling by the placement of the nerve upon the cooling device adds to the effects of cooling, so that block occurs at temperatures several degrees higher than would normally be the case (*Paintal* 1965).

Cooling, anodal polarization and local anaesthesia of the vagus nerves are effective in producing total reversible blockade ('reversible vagotomy'), which is much more satisfactory in reflex experiments than simply cutting the vagus nerves. However, no method is without limitations in its ability to block conduction in one set of vagal fibres without affecting the other.

5.3.1   Nerve Cooling

Cooling of nerve trunks is the most widely used and repeatable of differential blocking methods. Total block of conduction in myelinated fibres is achieved over a range of nerve temperatures between about 10° and 6°C; it occurs in a manner that is entirely independent of fibre diameter; the average blocking temperature for myelinated fibres is 6°–8°C (*Paintal* 1965; *Franz* and *Iggo* 1968; *Linden* et al. 1981; *Coleridge* et al. 1982b). Conduction in non-myelinated fibres continues at temperatures several degrees lower. Although the average blocking temperature for non-myelinated vagal fibres is given as roughly 3°C, many will conduct isolated impulses at 0°C (*Abbott* et al. 1965; *Franz* and *Iggo* 1968; *Coleridge* et al. 1982b).

In spite of a significant difference of several degrees in the temperatures at which A and C fibres are blocked completely, cooling affects conduction in both in a similar manner. Thus, beginning at temperatures only a few degrees below body temperature, the frequency of firing that traverses the cooled region of the nerve is progressively limited (*Franz* and *Iggo* 1968; *Linden* et al. 1981; *Coleridge* et al. 1982b). The limitation of impulse frequency imposed by nerve cooling is particularly apparent in non-myelinated fibres (*Franz* and *Iggo* 1968), perhaps because impulse transmission takes place over the whole of the axon membrane, involving significant changes in axonal Na^+ and K^+ concentrations and relatively high energy dissipation (*Douglas* and *Ritchie* 1962). Thus at nerve temperatures of 7°–8°C, conduction in C fibres is likely to be limited to discharge frequencies of 4–5 impulses s^{-1} or less, representing a substantial reduction of the C fibre response to most of the stimuli used in the experimental laboratory to evoke respiratory reflexes. For example, the response of bronchial C fibres to chemicals such as capsaicin, histamine and bradykinin typically consists of short bursts of impulses with instantaneous frequencies of 30–50 impulses s^{-1} (Figs. 5, 9, 11). It is likely that each of these bursts will be reduced to only one or two impulses when the vagus nerve is cooled to a temperature that will block conduction in myelinated fibres.

These progressive effects of cooling on the impulse frequency conducted across the block must be borne in mind in interpreting the results of reflex experiments. It has sometimes been assumed that the total contribution of C fibres to a particular respiratory reflex at 37°C is fully represented by the reflex effects that survive vagal cooling to 6°–8°C (*Karczewski* and *Widdicombe* 1969b, c). Judging from the effects of cooling on vagal reflexes that are known to be evoked solely by activation of lung C fibres, it appears that the selective blockade of A fibres at 6°–8°C may be associated with a 60% reduction in the reflex potency of afferent C fibres (*Roberts* et al. 1981b; *Coleridge* et al. 1982a).

5.3.2 Anodal Polarization

The technique of anodal polarization has been applied successfully to studies of respiratory reflexes in rabbits (*Guz* and *Trenchard* 1971; *Trenchard* et al. 1972; *Raybould* and *Russell* 1982). Current is applied to the nerve through bipolar electrodes with the anode placed centrally (*Casey* and *Blick* 1969; *Coleridge* et al. 1973b; *Whitwam* and *Kidd* 1975; *Thoren* et al. 1977); effects must be monitored at intervals by recording the compound action potential evoked by electrical stimulation across the blocked region, the polarizing current being adjusted to block the A wave but to leave the C wave intact. The current required to block conduction in A fibres varies widely with local conditions in the nerve.

One of the most troublesome complications of the method is extraneous stimulation at the cathode (*Casey* and *Blick* 1969; *Coleridge* et al. 1973b; *Whitwam* and *Kidd* 1975). *Thoren* et al. (1977), who recorded the effects of anodal polarization on the activity of single fibres, recommend cooling the nerve to 30°–32°C to avoid stimulation of C fibres. *Hopp* et al. (1980) have introduced a monopolar blocking technique which is equally successful in avoiding this problem. Anodal polarization appears to have one advantage over differential cooling: although there is some tendency for a limitation of C fibre discharge frequency when A fibres are blocked, it is probably not so marked as the limitation imposed by differential cold blockade. Hence anodal polarization is thought to provide a more realistic estimate of the contribution of afferent vagal C fibres to reflexes (*Thoren* et al. 1977). However, the current required to block conduction in A fibres decreases progressively with each successive application (*Whitwam* and *Kidd* 1975; *Thoren* et al. 1977), a finding interpreted as early evidence of nerve deterioration; moreover, the nerve becomes visibly damaged and discoloured with repeated applications (*Whitwam* and *Kidd* 1975; personal observations).

5.3.3 Local Anaesthesia

Local anaesthetics have been used in man (*Guz* et al. 1966, 1970) and in conscious dogs (*Phillipson* et al. 1970) to examine the effects on breathing pattern of totally blocking the vagus nerves, but they are used only rarely to produce differential block of nerve conduction. Susceptibility to local anaesthesia appears to be an inverse function of fibre diameter, the smaller fibres being blocked more readily as a consequence of their higher surface-to-volume ratio. Differential effects are monitored by recording the evoked compound action potential and adjusting the concentration of anaesthetics to one that will abolish the C wave but leave the A wave intact. The block is time dependent and is somewhat unreliable as a method of selectively blocking C fibres since small myelinated fibres, whose axon diameters may be close to those of C fibres, are often blocked after shorter exposures than those required to block C fibres (*Nathan* and *Sears* 1961; *Franz* and *Perry* 1974). Moreover, in the case of relatively thick nerve trunks, uneven penetration of anaesthetic makes the differential block less reliable (*Franz* and *Perry* 1974).

A quite different type of afferent blockade is produced when local anaesthetics are administered as aerosols to the lower respiratory tract, a method that has been applied to the study of lower airway reflexes in animals and man (*Jain* et al. 1973; *Dain* et al. 1975; *Cross* et al. 1976). In this case the anaesthetic penetrates from the mucosa and blocks afferent nerve endings unselectively. There is some doubt as to whether nerve bundles in the airway walls are also affected: some investigators report preservation of vagal efferent function (*Jain* et al. 1973), whereas others find that efferent conduction is abolished (*Dain* et al. 1975). The peripheral extent of local anaesthesia along the conducting airways is likely to be influenced by a number of factors, the most important being droplet size. With relatively large droplets (7–11 μm) the differential effect is upon reflexes: those elicited from the conducting airways are attenuated or abolished, whereas those elicited from the most distal lung divisions (e.g. the pulmonary chemoreflexes) are preserved and even accentuated (*Jain* et al. 1973).

5.4 Reflex Changes in Breathing

5.4.1 Effects of Stimulating Pulmonary C Fibres

The abrupt surge of pulmonary C fibre input evoked by intravenous, right atrial or pulmonary arterial injection of 5–20 μg kg^{-1} capsaicin in dogs and cats, or of 10–50 μg kg^{-1} phenyldiguanide in cats, causes an immediate apnoea in expiration (Figs. 5, 16) (*Dawes* et al. 1951; *Paintal* 1955; *Toh* et al. 1955; *Porszasz* et al. 1957; *Coleridge* et al. 1964b). The apnoea may persist for 30 s or more, but in spite of the resultant changes in blood gas tensions, when breathing resumes it is shallow, as well as rapid, the reduction in tidal volume being in marked contrast to the increase that might have been expected to result from the mounting stimulation of central and peripheral chemoreceptors during the period of apnoea. In cats, marked laryngeal constriction accompanies the apnoea if the dose of phenyldiguanide is large enough; the constriction gradually diminishes, with a brief phasic increase in each expiration, during the subsequent rapid shallow breathing (*Stransky* et al. 1973).

   Although an immediate arrest of breathing is often held to be the characteristic respiratory effect of stimulating pulmonary C fibres (J receptors), in fact the pattern of response varies with the dose of chemical administered, as *Brodie* (1900) first described. Rapid injection of large doses of phenyldiguanide in cats causes apnoea followed by rapid shallow breathing, whereas a slow rate of injection or injection of smaller doses causes rapid shallow breathing only (*Paintal* 1955; *Anand* and *Paintal* 1980). Presumably the latter method of administration avoids the brief intense surge of afferent input described by *Paintal* (1973) as contributing 'a strong artefactual element' to C fibre reflexes evoked by bolus injections. The prompt apnoea evoked in dogs by right atrial injection of capsaicin is often replaced by rapid shallow breathing when afferent C fibre input is attenuated by vagal cooling to 3°–5°C (*Coleridge* et al. 1964b). Halothane, which is also known to stimulate pulmonary C fibres (see above), has been shown to have respiratory effects that are dose dependent. Thus in dogs whose systemic circulation was artificially perfused, addition of 13% halothane to the inspired gas induced apnoea, whereas addition of 7% caused rapid shallow breathing only (*Lloyd* 1978). A similar though less pronounced dose-dependency is a feature of the respiratory response to stimulation of bronchial C fibres (see below).

In cats the apnoea involves not only a complete cessation of phrenic activity, but also inhibition of α and γ motor neurones to both inspiratory and expiratory intercostal muscles (*Schmidt* and *Wellhoner* 1970), and inhibition of both inspiratory and expiratory units in the medulla (*Koepchen* et al. 1977); hence the chest assumes its position of rest. This total, non-reciprocal inhibition of both inspiratory and expiratory units is in striking contrast to the reciprocal inhibition characteristic of the Hering-Breuer inflation and deflation reflexes (*Koepchen* et al. 1977). The pulmonary chemoreflex depression of expiratory units outlasts that of inspiratory units (*Schmidt* and *Wellhoner* 1970; *Koepchen* et al. 1977), and in the view of *Koepchen* et al. it is this imbalance that causes rapid shallow breathing.

Although apnoea (an arrest of breathing in expiration) is the typical immediate respiratory effect of the pulmonary chemoreflex evoked by large bolus injections in cats and dogs, and possibly also in man (*Jain* et al. 1972), it is not the typical response in rabbits. In rabbits, stimulation of pulmonary C fibres by phenyldiguanide causes tonic inspiration (resulting in an increased functional residual capacity) on which rapid shallow breathing movements are superimposed, and if breathing ceases it does so in inspiration (*Dawes* et al. 1951; *Karczewski* and *Widdicombe* 1969c; *Guz* and *Trenchard* 1971; *Miserocchi* et al. 1978), the overall pattern closely resembling that of Head's paradoxical reflex (*Widdicombe* 1967). Whatever the central mechanism of this inspiratory response, which appears to be peculiar to the rabbit, one may interpret its influence as beneficial, for rabbits, because of their frail chest walls, have such a small functional residual capacity at the position of rest (*Crosfill* and *Widdicombe* 1961) that prolonged apnoea in this position might lead to collapse of alveoli, requiring very large inspiratory efforts to reinflate the lungs.

An interesting hypothesis has been briefly formulated by *Trenchard* (1980) to account for the different chemoreflex respiratory responses in cats and rabbits. She postulates that phenyldiguanide stimulates two groups of lung C fibres, one exerting inhibitory effects on breathing, the other excitatory. If more 'inhibitory' endings were accessible from the pulmonary circulation and more 'excitatory' ones from the bronchial circulation, right atrial injection would evoke an initial apnoea followed, as the chemical reached the systemic circulation, by excitation of breathing: such might be the situation in cats. A reversal of the respective vascular accessibilities of 'inhibitory' and 'excitatory' afferents might account for responses like those described in rabbits. It is not easy to see how this hypothesis could be reconciled with the generally accepted view that the respiratory effects of stimulating pulmonary C fibres are dose dependent. Nevertheless, even the very notion of a dose-dependent respiratory response has been called into question (*Ginzel* 1978; *Lucas* and *Ginzel* 1980). Thus *Ginzel* (1978) injected phenyldiguanide into the right atrium and compared the resultant changes in spontaneous breathing movements in artificially ventilated cats with those in spontaneously breathing cats. He concluded that the respiratory effects were not dose dependent, that apnoea was the only primary respiratory response to stimulation of pulmonary C fibres and that rapid shallow breathing was due to secondary factors, such as decreased PaO_2 resulting from the apnoea, and could be prevented by appropriate measures. However, a full account of the experiments that support these provocative hypotheses (*Trenchard* 1980; *Ginzel* 1978) has yet to be published, and the weight of evidence still supports the conventional view that dose-dependency is a feature of the response to stimulation of pulmonary C fibres.

The central mechanism by which apnoea progresses to rapid shallow breathing is not fully understood, although stimulation of pulmonary C fibres is believed to cut short inspiratory (phrenic) activity (*Winning* and *Widdicombe* 1976; *Miserocchi* et al. 1978), and, if the initial stimulus is sufficiently intense, to cause an arrest of breathing by an extension of this effect (*Winning* and *Widdicombe* 1976).

Changes in inspiratory drive during the tachypnoeic phase of the pulmonary chemoreflex in cats have been inferred from changes in the impulse frequency of the phrenic bursts or in the V_T/T_I relationship. In some studies the development of inspiratory drive is reported to be unchanged, in others to decrease and in yet others to increase. *Winning* and *Widdicombe* (1976) and *Miserocchi* et al. (1978) found the V_T/T_I relationship to be unchanged during rapid shallow breathing and concluded that the development of inspiratory drive was unaltered. *Anand* and *Paintal* (1980) dissented from this view, finding the mean impulse frequency in the shortened phrenic bursts after phenyldiguanide to be less than during the control period, an observation taken to indicate a true inhibition of inspiratory—excitatory mechanisms by J receptors. A later brief account by *Anand* et al. (1982), however, presents the opposite view. Finding that right atrial injection of phenyldiguanide caused an increase rather than a decrease in mean phrenic impulse activity and, in some cats, a marked fall in end-tidal PCO_2 during the tachypnoeic phase, these investigators now

conclude that activation of J receptors causes both tachypnoea and an increased ventilatory drive. One may question the value of detailed analysis
of inspiratory patterns in non-steady-state responses such as the chemoreflex evoked by a bolus injection of phenyldiguanide. Phenyldiguanide is
known to stimulate arterial chemoreceptors (*Paintal* 1967), an effect that
may contribute to the breathing changes several seconds after the injection; moreover, secondary effects may result from the decrease in cardiac
output and from the changes in blood gas tensions that accompany the
initial apnoea. The most that can be said of the response under non-steady-
state conditions is that appropriate stimulation of pulmonary C fibres can
cause tachypnoea and a decrease in tidal volume.

In cats and dogs the rapid shallow breathing induced by injection of
phenyldiguanide or capsaicin may be prolonged for several minutes, thus
certainly outlasting the evoked discharge in pulmonary C fibres, which has
usually ended in less than 30 s (*Paintal* 1973; *Coleridge* and *Coleridge*
1977b), and probably also outlasting any effects on bronchial C fibres and
other susceptible afferents downstream to the pulmonary circulation. By
cutting the vagus nerves 5 s after injecting phenyldiguanide into the right
atrium, *Anand* and *Paintal* (1980) have shown that even a short-lived stimulation of pulmonary C fibres is sufficient to induce prolonged changes in
breathing.

The tachypnoea of the pulmonary chemoreflex has been interpreted
as a sensitization of the central inspiratory 'off-switch', with leftward dis-
placement of the Hering-Breuer threshold curve (*Winning* and *Widdicombe*
1976; *Miserocchi* et al. 1978). It does not follow, however, that a volume
signal from pulmonary stretch receptors is required. Thus *Miserocchi* et al.
(1978) evoked the pulmonary chemoreflex in cats by injecting phenyldiguanide, assessing the central timing of the respiratory response from
the phrenic neurogram, and found that rapid shallow breathing was independent of volume feedback, for both T_E and T_I were shortened even
when lung expansion was prevented by occluding the trachea in expiration.
Moreover, the tachypnoea evoked in rabbits by phenyldiguanide (*Guz* and
Trenchard 1971) and in dogs by capsaicin (*Coleridge* et al. 1964b) can still
be evoked when conduction in myelinated fibres is blocked by anodal
polarization or vagal cooling to 3°C, respectively. The tachypnoea of the
pulmonary chemoreflex, therefore, may be attributed to acceleration of
central respiratory rhythm.

The laryngeal constriction that may accompany the changes in breathing (*Stransky* et al. 1973), as well as the changes in breathing themselves,
are probably best interpreted as part of a protective response triggered by
access of potentially harmful agents to the distal divisions of the airways.

As described in an earlier section, reflex changes in breathing are
evoked in rabbits, dogs and cats when afferent C fibres are stimulated by

lung inflation (*Head* 1889; *Hammouda* and *Wilson* 1935a, b, 1939; *Widdicombe* 1954b). However, interpretation of the effects of forced lung inflation on breathing pattern is complicated by changes in blood gases resulting from interruption of normal gas exchange, and by the decrease in arterial pressure resulting from obstruction to venous return. Some of these problems were avoided in a recent study in which the effects of inflating the lungs and airways were examined in dogs on cardiopulmonary bypass (*Lloyd* 1979). The lungs were deprived of their normal pulmonary arterial blood supply, but lung gases were kept as close to normal as possible by intermittent ventilation with 5% CO_2 in O_2. Static inflation of a single lobe at pressures less than 20 cm H_2O decreased the frequency of diaphragmatic contractions, but when inflation pressure exceeded 20 cm H_2O effects were reversed, and frequency increased. The vagal blocking temperatures for these effects were not determined. It seems reasonable to assume, however, that the reversal of response at a critical static distending pressure resulted when the excitatory input in pulmonary C fibres overcame the inhibitory input of the low threshold pulmonary stretch receptors.

5.4.2 Effects of Stimulating Bronchial C Fibres

Bronchial C fibres have effects on the pattern of breathing generally similar to those of pulmonary C fibres. In artificially ventilated dogs, selective stimulation of bronchial C fibres by bronchial arterial injection of bradykinin briefly inhibited phrenic activity and then increased the frequency of the phrenic bursts and decreased their amplitude (*Coleridge* et al. 1981). Effects were often short-lived, probably because the stabilizing influence of the unchanged input from pulmonary stretch receptors caused the phrenic discharge to remain entrained to the ventilator cycle. *Coleridge* et al. (1983) therefore examined the effects of stimulating bronchial C fibres in spontaneously breathing dogs, in which the chest had been opened briefly to insert a catheter in the bronchial artery. A single injection of bradykinin (0.15–1.5 μg) into the right bronchial artery usually caused rapid shallow breathing (Fig. 18A); occasionally it evoked apnoea lasting 9–20 s, followed by variable changes in breathing (Fig. 18B). When bradykinin was infused slowly (0.2–2.0 μg min^{-1}, for 2–12 min), breathing invariably became rapid and shallow (Figs. 19, 20). Tachypnoea persisted until the end of the infusion and sometimes for several minutes afterwards. The changes in inspiratory pattern were usually confined to reduction of V_T without any change in V_T/T_I (*Coleridge*, *Coleridge* and *Roberts*, unpublished observations) and were therefore similar to those produced by stimulation of pulmonary C fibres (*Winning* and *Widdicombe* 1976; *Miserocchi* et al. 1978). In a few dogs, however, peak

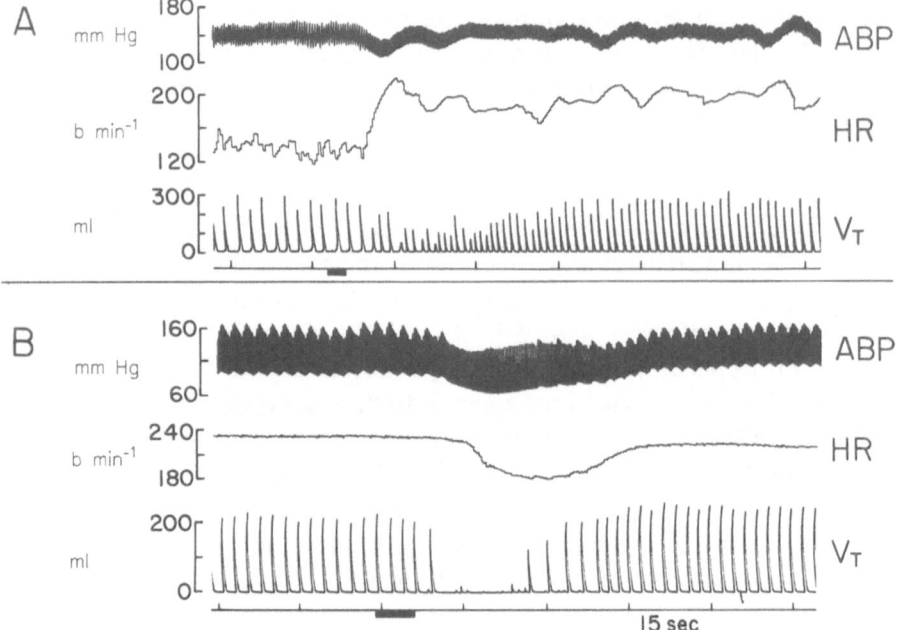

Fig. 18A, B. Effects on breathing of stimulating bronchial C fibres by injecting brady-kinin (**A** 0.5 μg; **B** 1.0 μg) into the right bronchial artery (injection signalled by *black bar* on time trace). **A** and **B** in different dogs. Note rapid, shallow breathing in **A** and apnoea in **B**; note cardiac acceleration in **A** and slowing in **B**. *ABP*, arterial blood pressure; *HR*, heart rate; V_T, tidal volume. (*Coleridge* et al. 1983)

inspiratory airflow, and hence V_T/T_I, increased. The rapid shallow breathing had variable effects on end-tidal PCO_2, but in nearly half the experiments the increased frequency more than compensated for the decreased tidal volume, and the resulting hyperventilation decreased end-tidal PCO_2 by 2–9 mm Hg for the duration of the infusion, indicating that bronchial C fibres had an excitatory influence on respiratory control (*Coleridge* et al. 1983).

The prolonged tachypnoea evoked by bronchial arterial infusion of bradykinin was abolished by cutting the vagus nerves or by cooling them to 0°C, and in the latter case was restored by rewarming the nerves (*Coleridge* et al. 1983). The respiratory response to bradykinin was not altered by indomethacin; hence it could be attributed to a direct action of brady-kinin on bronchial C fibres, uncomplicated by prostaglandin release (*Coleridge* et al. 1983).

Residual effects, usually consisting of irregular spasmodic gasps, were sometimes evoked by bradykinin after vagotomy; they were abolished by avulsing the upper thoracic sympathetic chain and rami. Similar residual effects observed in vagotomized dogs and cats when irritant gases were administered to the lower trachea were also abolished by interrupting upper

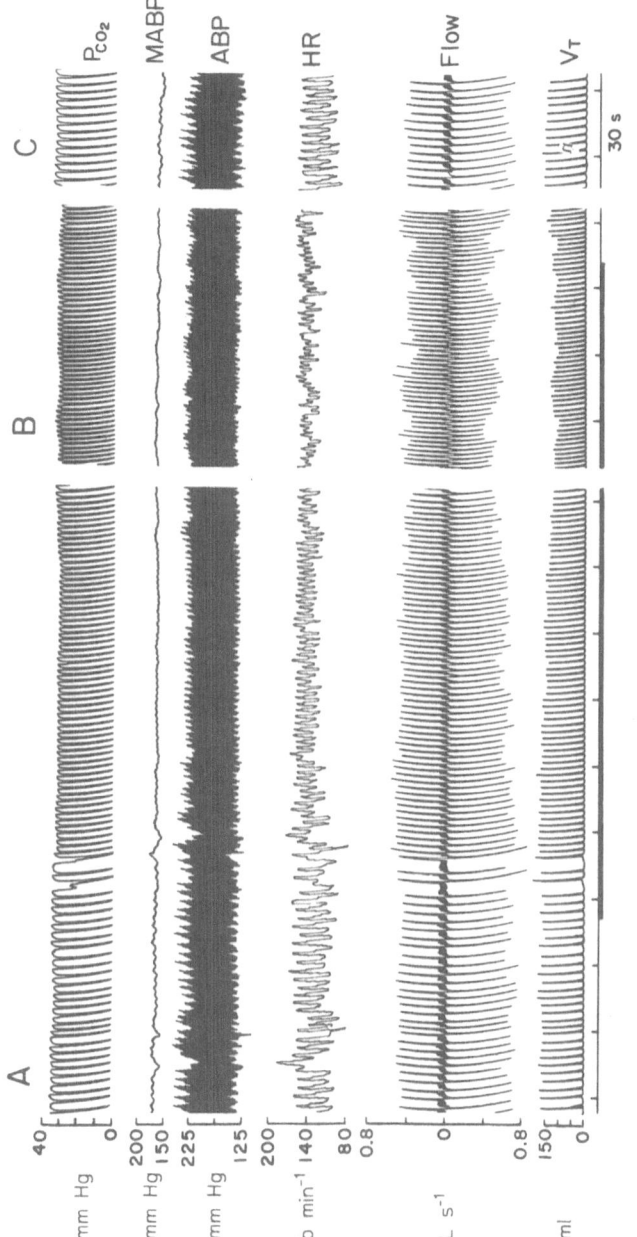

Fig. 19. Ventilatory effects of stimulating bronchial C fibres; changes in the pattern of breathing produced by prolonged (12 min) infusion of bradykinin (0.5 μg min^{-1}) into the right bronchial artery of a dog. Infusion (signalled by *black bar* on time trace) began in *A* and continued without interruption to end in *B*; interval of about 7 min between *A* and *B* and of 4 min between *B* and *C*. Note that mean arterial blood pressure was virtually unchanged and that end-tidal PCO_2 decreased. PCO_2, tidal PCO_2; *MABP*, mean arterial blood pressure; *ABP*, arterial blood pressure; *HR*, heart rate; *Flow*, airflow; V_T, tidal volume. (*Coleridge* et al. 1983)

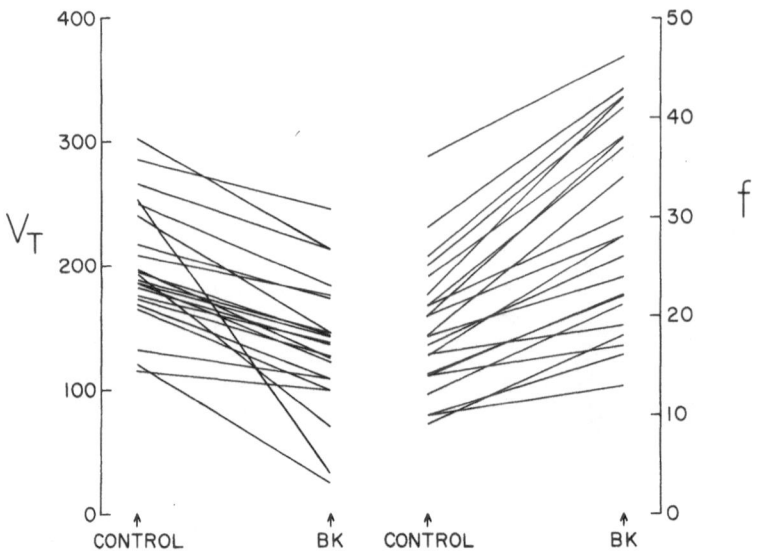

Fig. 20. Changes in tidal volume and breathing frequency evoked in 22 experiments on ten dogs by infusing bradykinin 0.2−2.0 μg min^{-1} for 2−12 min into the right bronchial artery to stimulate bronchial C fibres. Control values were measured as the average over 30 s immediately before the infusion; values after bradykinin *(BK)* were averaged over 30 s at the peak of the response. V_T, tidal volume (ml); *f*, breathing frequency (breaths min^{-1}). (Based on results of experiments described by *Coleridge* et al. 1983)

thoracic sympathetic pathways (*Cromer* et al. 1933; *Banister* et al. 1950; *Widdicombe* 1954c).

   Administration of aerosols of 0.1% bradykinin to the lower airways also evoked rapid shallow breathing in dogs by stimulating airway C fibres selectively, including C fibres whose endings were in the lower trachea (*Coleridge* et al. 1983). In some dogs, the ventilatory effects of aerosol inhalation were similar to those of bronchial arterial infusion and consisted of prolonged and regular rapid shallow breathing. In other dogs, breathing became so rapid, shallow and irregular that its rate and depth were not measurable (Fig. 21). When vagal conduction was intact, the effects of bradykinin aerosol had a latency of 18−45 s and continued for 1−4 min after the aerosol ended. By contrast, when the vagus nerves were cooled to 0°C, inhalation of aerosol had no effect at all, even when continued for as long as 7 min; however, the response was immediately restored by rewarming the nerves. The experiment depicted in Fig. 21 illustrates the powerful responses triggered by bradykinin; it also emphasizes the central role of afferent vagal input in these responses.

   These experiments were not the first to show that reflex changes in breathing can be evoked by stimulation of vagal afferents that receive

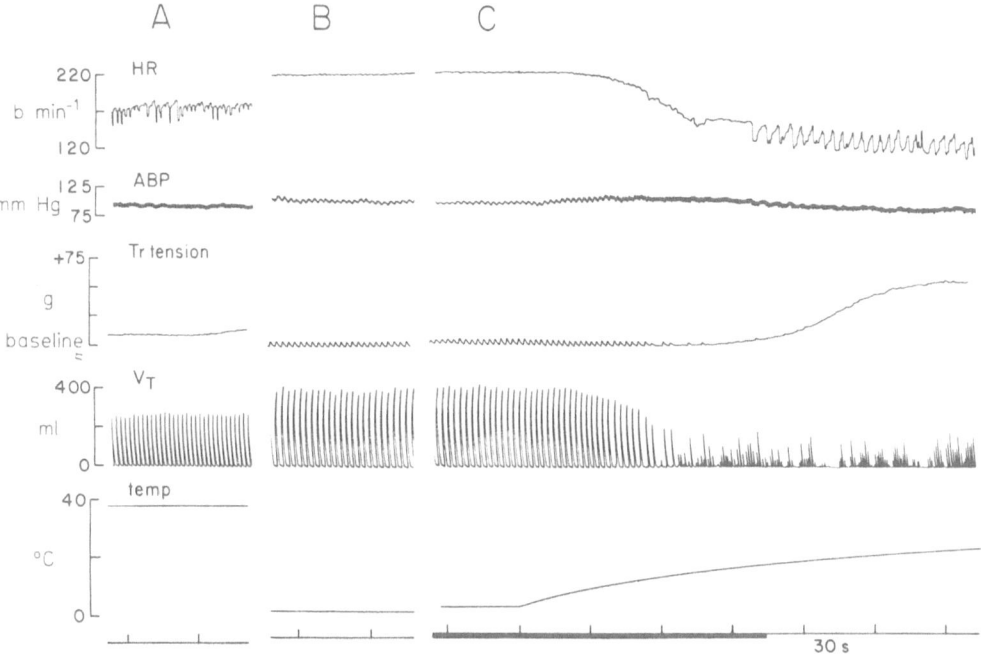

Fig. 21. Changes in breathing, heart rate and tracheal smooth muscle tone evoked in a dog by delivery of bradykinin aerosol to the lower trachea; effects delayed by vagal cooling. *A*, control, vagal temperature 37°C. Between *A* and *B*, the vagus nerves were cooled to 0°C (they were not rewarmed until *C*). *B*, After 5 min of vagal cooling; note decrease in breathing frequency and increase in tidal volume and heart rate. Between *B* and *C*, administration of bradykinin aerosol (0.1% solution) began; it continued for 7 min (signalled by *black bar* in *C*). *C*, Vagus nerves rewarmed and aerosol terminated. Rapid shallow breathing, tracheal contraction and cardiac slowing continued for 5–10 min after the end of *C*. *HR*, heart rate; *ABP*, arterial blood pressure; *Tr tension*, tracheal tension in grams above a baseline set at 75 g; V_T, tidal volume; *temp*, temperature of vagus nerves. (*Coleridge* et al. 1983)

their blood supply from the bronchial circulation. Rapid shallow breathing, sometimes preceded by apnoea, was oberved by *DeKock* et al. (1966) when they injected histamine into the bronchial artery in dogs; these effects were preceded or interrupted by an isolated deep breath or sigh. It seems likely that this combination of effects was triggered by the stimulation of more than one type of lung afferent. Histamine stimulates bronchial C fibres (*Coleridge* and *Coleridge* 1977b; *Coleridge* et al. 1978a), as well as irritant receptors (*Mills* et al. 1969; *Sampson* and *Vidruk* 1975). It seems likely that the apnoea and rapid shallow breathing reported by *DeKock* et al. (1966) resulted from stimulation of airway C fibres; the gasps or sighs were probably a manifestation of the 'gasp reflex' evoked by stimulation of irritant receptors (*Sellick* and *Widdicombe* 1970; *Glogowska* et al. 1972).

Airway C fibres may play some part in initiating the small and variable tachypnoea evoked by gross inflation of the extrapulmonary airways in dogs on cardiopulmonry bypass (*Lloyd* 1979), but the possibility has yet to be confirmed by selective blockade of myelinated pathways in the vagus.

5.5   Reflex Effects on Airway Smooth Muscle

5.5.1   *Introduction*

Although smooth muscle is present in all the airways from the extrathoracic trachea to the alveolar ducts, nervous regulation affects mainly the smooth muscle of the larger airways with cartilage in their walls, including that of the bronchi of intermediate size, whose calibre is the major determinant of airflow resistance in the lungs; the smooth muscle of the smaller airways, whose tone affects lung compliance rather than airflow resistance, does not in general appear to be under nervous control (*Woolcock* et al. 1969; *Karczewski* and *Widdicombe* 1969a; *Stephens* and *Kroeger* 1980; *Nadel* 1980; *Orehek* 1981). Like the smooth muscle of the gut, airway smooth muscle has a dual efferent nerve supply: parasympathetic and sympathetic. The parasympathetic (vagus) nerves are excitatory to airway smooth muscle and cause bronchoconstriction. The role of the sympathetic innervation is controversial and varies with the species (*Orehek* 1981). Unlike the smooth muscle of the gut, airway smooth muscle has little or no intrinsic tone; however, tonic low-frequency activity in vagal bronchomotor fibres (*Widdicombe* 1966) appears to maintain a significant degree of airway smooth muscle tone under normal conditions, and airflow resistance is decreased by cutting or cooling the vagus nerves or by administering atropine (*Nadel* 1980).

The vagal centres controlling the tone of airway smooth muscle receive a drive from the medullary chemoreceptors (*Loofbourrow* et al. 1957; *Nadel* and *Widdicombe* 1962a); they also receive input from arterial chemoreceptors and baroreceptors (*Nadel* and *Widdicombe* 1962a) and from chemosensitive endings in skeletal muscle (*Coleridge* et al. 1982a; *Kaufman* et al. 1982b), as well as from afferent nerve endings in the respiratory tract itself. In regard to the reflex role of input from the upper respiratory tract, mechanical and chemical stimuli applied to the nose have inhibitory effects on airway smooth muscle (*Tomori* and *Widdicombe* 1969; *Allison* et al. 1974), while those applied to the larynx have excitatory effects (*Nadel* and *Widdicombe* 1962b; *Boushey* et al. 1972). As to the role of input from the lower airways, slowly adapting stretch receptors have a tonic inhibitory influence on airway smooth muscle (*Widdicombe* and *Nadel* 1963), whereas chemically sensitive nerve endings have an excitatory one. The identity of these excitatory, chemically sensitive nerve endings has aroused much interest, particularly in view of the possibility that reflex bronchoconstriction is triggered or aggravated by release of lung autocoids in airway diseases and by the action of chemical air pollutants in the intrathoracic airways. Traditionally, the reflex bronchoconstriction evoked by administration of histamine to the lower airways has been attributed to stimulation of rapidly adapting (irritant) receptors with myelinated fibres (*Mills* et al. 1969; *Gold* et al. 1972; *Karczewski* and *Widdicombe* 1969c; *Nadel* 1980; *Widdicombe* 1974a, b, 1977a, b, 1981), and until recently an exclusive role for irritant receptors has been postulated in bronchoconstrictor reflexes evoked by a variety of lung autocoids and foreign airway irritants (*Widdicombe* 1974a, b, 1977a, b). There is now increasing evidence that non-myelinated afferents from the lower respiratory tract play an important and previously unsuspected role in such reflexes, and that low-frequency background activity in these afferent C fibres exerts a tonic excitatory influence on vagal bronchoconstrictor tone.

Reflex changes in bronchomotor tone have been studied by measuring lung resistance (*Nadel* and *Widdicombe* 1962a), by measuring airway diameters in serial tantalum bronchograms (Fig. 22) (*Nadel* et al. 1968; *Russell* and *Lai-Fook* 1979) and by recording isometric contraction in the trachealis muscle of an innervated segment of the upper trachea (Fig. 23) (*Brown* et al. 1980; *Roberts* et al. 1981b). The last has certain advantages in reflex studies: it permits investigation of an airway smooth muscle whose ultrastructural, mechanical and pharmacological properties are generally similar to those of smooth muscle in large bronchi (*Stephens* and *Kroeger* 1980) but which is more accessible; it allows variations in airway smooth muscle tone to be recorded continuously; and it also allows reflex bronchomotor and bronchosecretory mechanisms to be studied simultaneously in a single preparation. Moreover, since the upper trachea receives most of its motor innervation from the superior laryngeal nerves, the recurrent and pararecurrent nerves can be cut, separating the afferent and efferent limbs of a vagal reflex originating in the lungs. The investigator is then able to examine the effects of blocking the afferent pathway in the cervical vagal trunk without interrupting the vagal efferent pathway to the smooth muscle of the tracheal segment (Figs. 23, 25).

Tracheal and bronchial smooth muscles have similar length–tension characteristics in vitro (*Russell* 1978; *Souhrada* and *Dickey* 1976) and their responses to chemical agonists appear comparable (*Hendrix* et al. 1983). Even so, it is possible that elicitation of reflex isometric tracheal contraction is not necessarily accompanied by a physiologically significant change in pulmonary resistance (*Leff* et al. 1982). However, stimulation of pulmonary C fibres in dogs by a similar dose of capsaicin evokes reflex contraction of the tracheal segment (*Coleridge* et al. 1982a) and a reflex decrease of 20% in peripheral airway diameter measured by tantalum bronchograms (*Russell* and *Lai-Fook* 1979), effects on trachea and peripheral airways having a similar time course. Hence reflex changes in tracheal tone appear to provide a good indication of the direction of changes in tone in more peripheral airways.

5.5.2 Role of Pulmonary C Fibres

That bronchoconstriction is a component of the pulmonary chemoreflex was suggested by the observations of *Barer* and *Nusser* (1953) in cats. Clear evidence of bronchoconstriction was obtained by *Karczewski* and *Widdicombe* (1969c), who found that right atrial injection of phenyl-diguanide evoked a reflex increase in airway resistance in rabbits. The latter

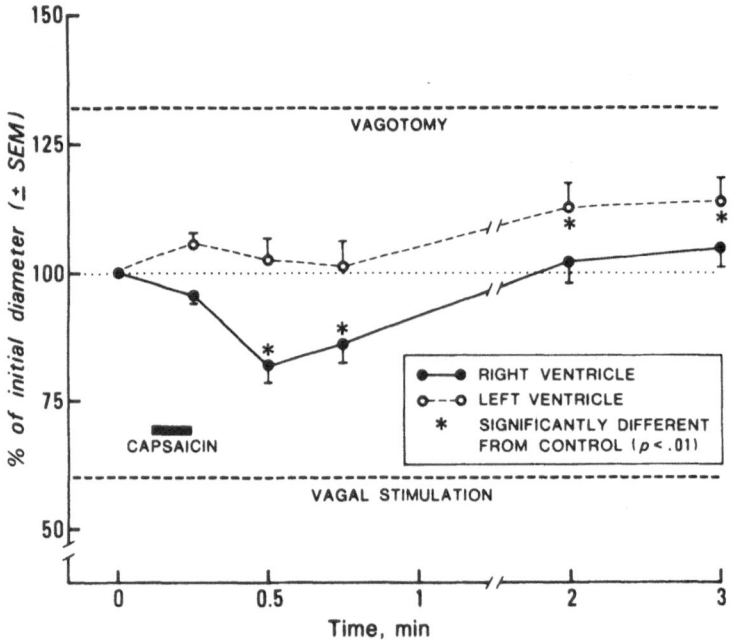

Fig. 22. Comparison of the effects of right and left ventricular injection of capsaicin on airway diameters in artificially ventilated dogs, as revealed by tantalum broncho-grams. Intrapulmonary airways were outlined with tantalum dust and diameters were determined by taking serial roentgenograms at end-expiration. Capsaicin (20 μg kg^{-1}) was injected during a 7-s period indicated by the *horizontal solid bar*. Each point re-presents the mean diameter of 36 airways (6 dogs) expressed as a percentage of the control diameter (time 0). The *horizontal dashed lines* indicate: *above,* effects of bilateral vagotomy; *below,* effects of bilateral stimulation of the peripheral ends of the cut vagus nerves (25 V, 25 Hz, 3 ms pulses) to produce maximal bronchoconstriction. (*Russell* and *Lai-Fook* 1979)

investigators also observed that a small reflex component of the increase in airway resistance evoked by pulmonary anaphylaxis in rabbits was still present after the vagus nerves had been cooled to 8°–10°C, and they concluded that 'deflation receptors' supplied by C fibres (i.e. pulmonary C fibres) contributed to these bronchoconstrictor effects (*Karczewski* and *Widdicombe* 1969b). Subsequently *Russell* and *Lai-Fook* (1979) demonstrated by means of tantalum bronchograms that injection of capsaicin into the right ventricle of dogs caused reflex bronchoconstriction (Fig. 22). Since airway diameter was largely unaffected by injection of capsaicin into

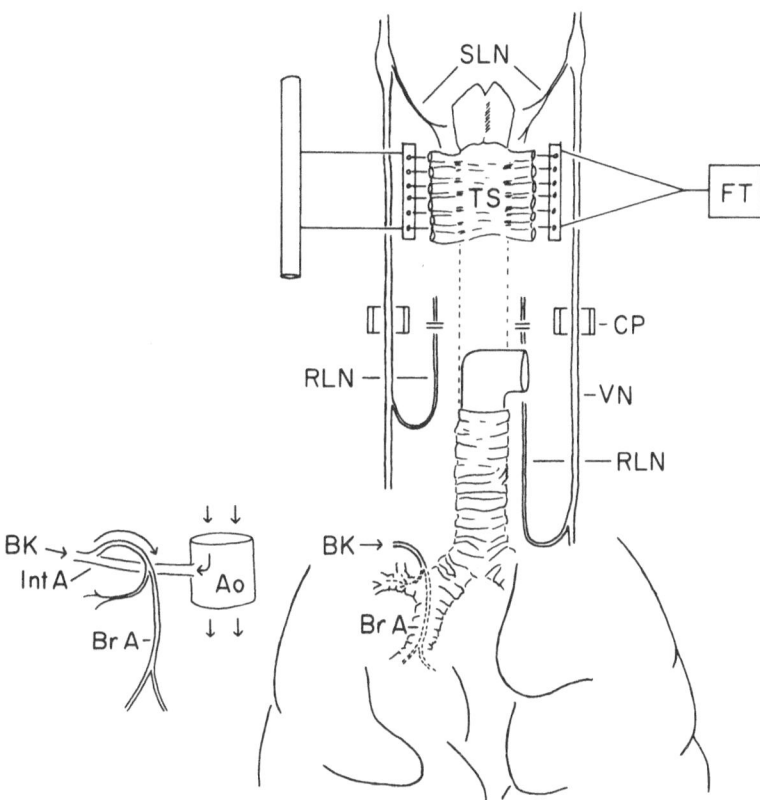

Fig. 23. Innervated tracheal segment *(TS)* for examining the reflex control of airway smooth muscle in dogs. The upper trachea is incised ventrally in mid-line, and each cut edge is retracted laterally and attached to a light plastic bar, one bar being anchored to a fixed metal rod, the other attached to an isometric force transducer *(FT)* mounted on a rack and pinion. The segment is stretched initially to a baseline tension of 50 or 75 g. Recurrent (and pararecurrent) laryngeal nerves *(RLN)* are cut so that segment is innervated only by superior laryngeal nerves *(SLN)*. Cervical vagus nerves *(VN)* are placed on cooling platforms *(CP)*. Right bronchial artery *(Br A)* lies dorsal *(dotted line)* to right lung root. *Smaller diagram on left* depicts right bronchial artery stemming from an intercostal artery *(Int A)*, which arises from thoracic aorta *(Ao)*. Bradykinin *(BK)* or other chemicals injected retrogradely into intercostal artery pass into bronchial artery. (*Roberts* et al. 1981b)

the left ventricle, and since the effects of right ventricular injection were abolished by vagotomy, they reasonably attributed the bronchoconstriction to a vagal reflex initiated by stimulation of pulmonary C fibres.

The reflex influence of pulmonary C fibres on airway smooth muscle in dogs was also demonstrated by *Coleridge* et al. (1982a), using the innervated tracheal segment. Injection of capsaicin into the right atrium usually evoked an increase in tracheal tension, whereas injection into the left atrium usually did not (Fig. 24). The contraction evoked by right atrial injection still occurred after conduction in myelinated fibres had been blocked by cooling the mid-cervical vagus nerves to 6–7°C but was abolished by cutting the nerves or by cooling them to 0°C. Since the tracheal segment was supplied only by the superior laryngeal nerves (see above), the response to right atrial injection was due to a reflex whose afferent arm was in the vagus nerve, and since pulmonary C fibres are the only vagal afferents in dogs to be stimulated by capsaicin reaching them from the pulmonary circulation (*Coleridge* et al. 1965; *Coleridge* and *Coleridge* 1977b), clearly the reflex was triggered by them.

Reflex contraction of tracheal smooth muscle is also evoked in dogs when pulmonary C fibres are stimulated by inflating the lung (*Roberts* et al. 1972a). In this case effects on airway smooth muscle are somewhat analogous to the paradoxical effects on breathing described by *Head* (1889) and *Widdicombe* (1967), in that they are not clearly revealed until the vagus nerves are cooled sufficiently to abolish the overriding inhibitory influence of slowly adapting pulmonary stretch receptors on the airways (*Widdicombe* and *Nadel* 1963). When the vagus nerves are at body temperature, lung inflations of 2–3 V_T above functional residual capacity evoke relaxation of the tracheal segment (*Roberts* et al. 1982a). When the vagus nerves are cooled, however, the inflation-evoked relaxation is gradually reduced and at vagal temperatures of between 10°C and 6°C inflation evokes contraction. With further vagal cooling, the inflation-evoked contraction is progressively reduced and is abolished at 0°–2°C. Rewarming the vagus to body temperature restores the original inhibitory response. Since the afferent input from slowly adapting and rapidly adapting stretch receptors is blocked at 6°–7°C, whereas that in C fibres is not completely blocked until 0°C (*Coleridge* et al. 1982b), the inflation-evoked contracttion unmasked by vagal cooling can be attributed mainly to stimulation of pulmonary C fibres. The contribution of bronchial C fibres to the increase in tracheal tension is probably small, because they are in general only weakly stimulated by the degree of lung inflation (2–3 V_T) used in these experiments (*Coleridge* and *Coleridge* 1977b; *Kaufman* et al. 1982a).

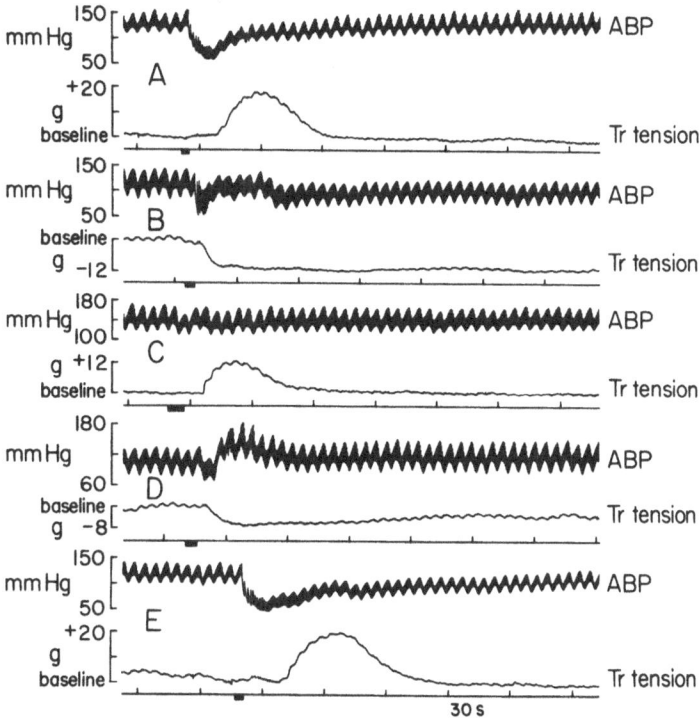

Fig. 24A–E. Changes in tracheal tension and arterial blood pressure evoked in a dog by injecting capsaicin at different sites in the bloodstream: **A** right atrium ($10 \mu g \ kg^{-1}$); **B** left atrium ($10 \mu g \ kg^{-1}$); **C** right bronchial artery ($3 \mu g$); **D** femoral artery ($50 \mu g$); **E** femoral vein ($10 \mu g \ kg^{-1}$). Tracheal contraction in **A** and **E** was due to stimulation of pulmonary C fibres, and that in **C** to stimulation of bronchial C fibres; tracheal relaxation in **B** and **D** was due to stimulation of afferent nerves in skeletal muscle. *ABP*, arterial blood pressure; *Tr tension*, tracheal tension (baseline tension set at 75 g). (*Coleridge* et al. 1982a)

5.5.3 Role of Bronchial C Fibres

Russell and *Lai-Fook* (1979) found little or no reduction in airway diameter in dogs when they injected capsaicin into the left heart, though such an injection is known to stimulate bronchial C fibres (*Coleridge* and *Coleridge* 1977b). Indeed, *Russell* and *Lai-Fook* observed that left ventricular injection even evoked slight bronchodilatation (Fig. 22). These results could be taken to indicate that bronchial C fibres had little bronchoconstrictor action, at least in dogs. *Russell* and *Lai-Fook* were careful to point out, however, that their findings did not exclude a reflex action of bronchial C fibres on airway smooth muscle, because opposing reflexes set in train by the effect of capsaicin on other susceptible afferent endings, located in the heart and great vessels (*Coleridge* et al. 1964a, 1973a) or elsewhere, may have masked the effects of stimulating bronchial C fibres.

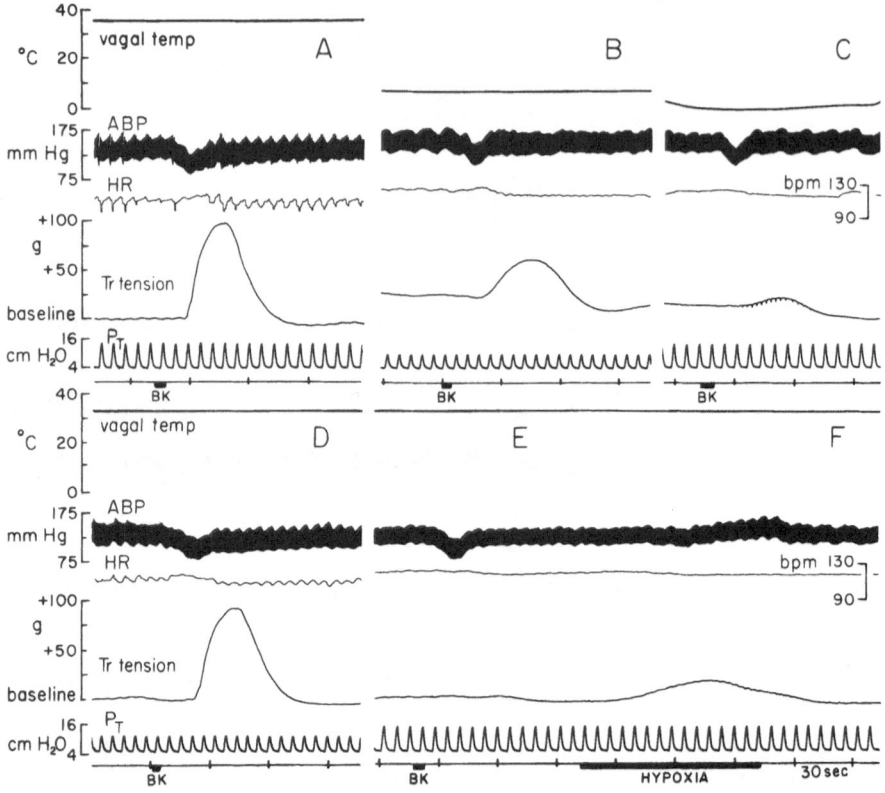

Fig. 25. Reflex contraction of tracheal smooth muscle evoked by stimulating bronchial C fibres, and abolition of the response by cooling or cutting mid-cervical vagus nerves. Dog, chest open and lungs ventilated by a pump; recurrent and pararecurrent nerves were cut. *A–E*, 1.5 μg bradykinin *(BK)* injected into right bronchial artery: *A*, vagal temperature 36°C; *B*, vagi cooled to 7°C (note increase in baseline tension); *C*, vagi cooled to 0°–1°C; *D*, vagi rewarmed. Vagi cut between *D* and *E*. *E*, response to bradykinin abolished but reflex tracheal contraction could still be evoked by ventilating lungs with 5% O_2 in N_2 *(F)* (at all other times in *A–F*, lungs ventilated with 50% O_2 in air). Note changes in tracheal pressure; lungs were briefly hyperinflated between *A* and *B*, and between *C* and *D*. *Vagal temp*, temperature of mid-cervical vagus nerves on cooling platforms; *ABP*, arterial blood pressure; *HR*, heart rate; *Tr tension*, tracheal tension (baseline, 50 g); P_T, tracheal pressure. (*Roberts* et al. 1981b)

This suggestion has proved to be correct, for selective stimulation of bronchial C fibres by injection of bradykinin directly into the bronchial artery evoked an unequivocal reflex increase in tracheal smooth muscle tension (Fig. 25) (*Roberts* et al. 1981b). Sensitivity to bradykinin varied widely from dog to dog but in a given dog the response was dose dependent. Tracheal tension increased in all dogs after injection of 1.5 μg bradykinin; in half the dogs one-tenth of this dose evoked contraction, and in one dog as little as 19 ng was effective. Contraction could still be evoked when myelinated vagal fibres were blocked by cooling to 7°C but was abolished when non-myelinated fibres were blocked by cooling to 0°–1°C (Fig. 25).

The afferents responsible for tracheal contraction were undoubtedly situated within the immediate vascular territory supplied by the bronchial artery, for injection of similar doses of bradykinin into the left atrium had little effect on tracheal tension (*Roberts* et al. 1981b). Bradykinin has no direct effect on airway smooth muscle in dogs (*Waaler* 1961; *Roberts* et al. 1981b). Since bronchial C fibres were the only lung afferents to be stimulated significantly by bronchial arterial injection of such small amounts of bradykinin (*Kaufman* et al. 1980b; *Roberts* et al. 1981b), there is no doubt that the reflex tracheal contraction was initiated by bronchial C fibres.

Stimulation of bronchial C fibres by capsaicin caused reflex tracheal contraction when doses one hundredth or less of those that failed to trigger contraction from the left atrium were injected directly into the bronchial artery (Fig. 24) (*Coleridge* et al. 1982a). Effects were dose dependent and contraction could be evoked by as little as 150 ng, a dose that had no effect when injected downstream to the bronchial vascular bed. Effects survived vagal cooling to 6°–8°C but were abolished by cooling to 0°C. Action potential studies confirmed that bronchial C fibres were the only intrapulmonary afferents to be stimulated by the small doses of capsaicin injected in the reflex experiments. Serial injection of capsaicin at more peripheral sites revealed that afferents stimulated by left atrial injection and capable of masking the bronchoconstrictor action of bronchial C fibres were located in hindlimb skeletal muscle (Fig. 24D) (*Coleridge* et al. 1982a; *Kaufman* et al. 1982b). There may be as yet unidentified afferents at other sites with similar reflex bronchodilator properties.

Reflex tracheal contraction was also evoked by inhalation of bradykinin aerosol (0.01%–0.1%) through a cannula in the lower trachea (Figs. 21, 26) (*Coleridge* et al. 1983). The response, which was abolished by cutting or cooling the vagus nerves, appeared to be due solely to stimulation of afferent vagal endings in the lower respiratory tract, for the absence of any change in blood pressure was good evidence that bradykinin had not entered the general bloodstream (Figs. 21, 26). (Bradykinin is a powerful vasodilator, acting directly on vascular smooth muscle, and infusion of as little as 0.5 μg min^{-1} into the aorta decreased blood pressure but had no effect on tracheal tone.)

In all the above experiments in which bronchial C fibres evoked reflex contraction of tracheal smooth muscle, the inhibitory influence of slowly adapting pulmonary stretch receptors on bronchomotor tone remained relatively constant because the lungs were ventilated artificially. In spontaneously breathing animals, however, airway defence reflexes are characterized by tachypnoea. Under these conditions the question arises whether, in spite of the reduced tidal volume, pulmonary stretch receptor input is increased sufficiently by tachypnoea to oppose effectively the broncho-

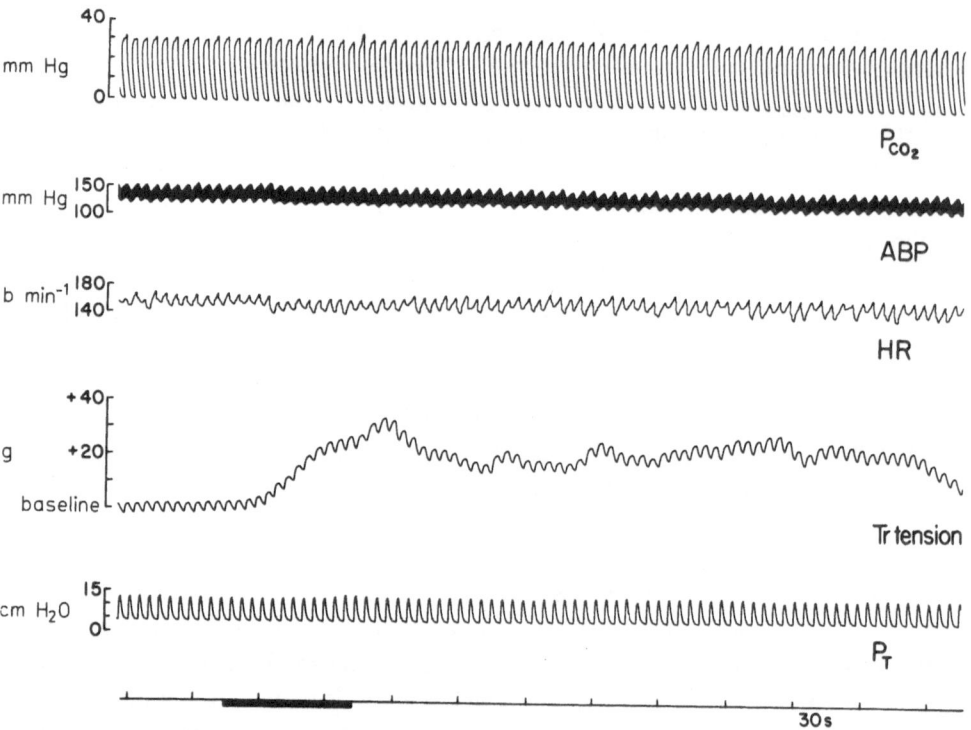

Fig. 26. Stimulation of airway C fibres by administration of bradykinin aerosol (0.01% solution) to the lower trachea in a dog evokes reflex contraction of an upper tracheal segment (aerosol delivery signalled by *black bar*). The chest was open and the lungs were ventilated by positive pressure. The tracheal response was subsequently abolished by cutting the vagus nerves. PCO_2, tidal CO_2; *ABP*, arterial blood pressure; *HR*, heart rate; *Tr tension*, tracheal tension in grams above a baseline set at 75 g; P_T, pressure in lower trachea. (*Coleridge, Coleridge* and *Roberts*, unpublished)

constrictor influence of bronchial C fibres. In the event, bradykinin aerosols administered to the lower airways of spontaneously breathing dogs caused sustained tracheal contraction in spite of an often marked degree of tachypnoea (*Coleridge* et al. 1983).

5.5.4 Role of C Fibres in Bronchomotor Tone

There is evidence that vagal bronchomotor tone in cats is entirely dependent upon vagal input, for *Jammes* and *Mei* (1979) found that selective section of afferent vagal fibres at the level of the nodose ganglion (*Mei* 1966) appeared to abolish all bronchoconstrictor tone. Finding that bronchomotor tone was also abolished when procaine was applied to the intact vagus nerves in a concentration that selectively blocked the C-wave of the evoked compound action potential, *Jammes* and *Mei* further postulated that vagal bronchoconstrictor tone is maintained largely by background activity in C fibres from the lung. These observations are open to the objection that the application of local anaesthetic must have blocked efferent as well as afferent vagal pathways, because not only are many vagal preganglionic fibres non-myelinated (*Agostoni* et al. 1957) but local anaesthetic block does not discriminate adequately between non-myelinated and small myelinated fibres (*Nathan* and *Sears* 1961; *Franz* and *Perry* 1974). Nevertheless, the hypothesis of *Jammes* and *Mei* (1979) receives some support from the observation of *Roberts* et al. (1981b, 1982a) in dogs that progressive cooling of the lower cervical vagus nerves increased smooth muscle tension in an upper tracheal segment innervated only by the superior laryngeal nerves. Tension reached a peak at 7°–8°C and then decreased as the temperature was reduced to 2°C. The initial increase in tone was attributed to abolition of the inhibitory influence of slowly adapting pulmonary stretch receptors (*Widdicombe* and *Nadel* 1963) and to unmasking of the excitatory influence of afferent C fibres, which were themselves blocked by further cooling. Since this excitatory effect could no longer be obtained after the pulmonary vagal branches had been cut, it was probably initiated by background activity in lung C fibres. These results (*Roberts* et al. 1981b, 1982a) provide support for the hypothesis that input from lung C fibres contributes to bronchomotor tone, but they do not justify the conclusion that lung C fibres are solely responsible. *Jammes* and *Mei*'s results appear to discount the possible influence of other mechanisms on vagal bronchoconstrictor tone, but the results of some investigators indicate that the central action of CO_2 is a powerful factor (*Loofbourrow* et al. 1957; *Widdicombe* 1966; *Richardson* et al. 1982). We have found that in some dogs even a small decrease in end-tidal PCO_2 is sufficient to decrease tracheal smooth muscle tension, an affect that is abolished by section of the superior laryngeal nerves (*Coleridge, Coleridge* and *Roberts*, unpublished observations).

5.6 Reflex Changes in Tracheobronchial Secretion

The formation of mucus by the submucosal glands and surface epithelial goblet and serous cells of the tracheobronchial tree is an integral part of the defence mechanisms of the lungs and airways (*Reid* 1960; *Widdicombe* 1978; *Nadel* et al. 1979; *Nadel* and *Davis* 1980). Mucus acts as a barrier to penetration by noxious chemicals and physical agents, absorbing or trapping them and limiting their passage into the tissues; it also serves as the vehicle in which these noxious agents are moved up the airways by the sweeping action of the cilia, to be swallowed or expelled by coughing. In dogs, cats and humans, submucosal glands and epithelial goblet and serous cells are present in airways that contain cartilage (i.e. in trachea and larger bronchi but not in bronchioles). Submucosal gland secretion is increased by stimulation of parasympathetic nerves and by cholinergic drugs, as well as by stimulation of sympathetic adrenergic nerves (*Gallagher* et al. 1975; *Nadel* and *Davis* 1980). Airway secretion is increased reflexly by stimulation of carotid body chemoreceptors (*Davis* et al. 1982a) and afferent endings in the respiratory tract (*Phipps* and *Richardson* 1976; *Widdicombe* 1978; *Nadel* et al. 1979; *Nadel* and *Davis* 1980; *Davis* et al. 1982b).

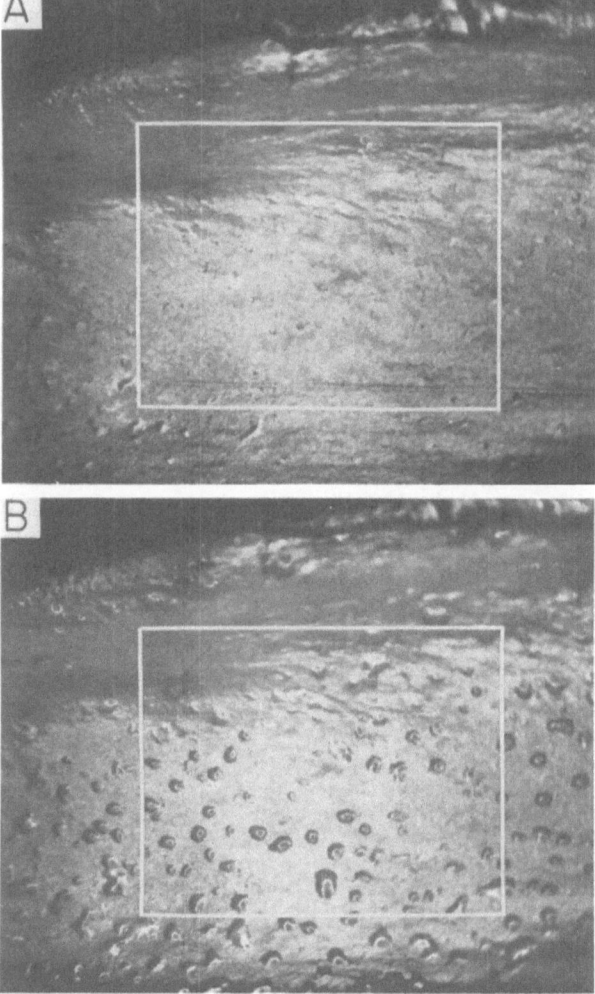

Fig. 27A, B. Photomicrographs of tantalum-coated mucosa on lateral wall of upper tracheal segment (viewed through a dissecting microscope). White lines enclose area of 1.2 cm^2, which was displayed on television screen; the image was also recorded on videotape for playback and measurement of the rate of secretion. **A** Appearance of mucosa 60 s after it had been dried and sprayed with powdered tantalum, i.e. at end of control period and immediately before 1.5 μg bradykinin was injected into bronchial artery. **B** Appearance of mucosa 60 s after injection of bradykinin. (*Davis* et al. 1982b)

Investigators have employed two methods to examine the reflex control of airway secretion in cats and dogs, the activity of submucosal glands in the more accessible trachea being taken as an index of secretion in peripheral airways. In cats, the glycoprotein fraction of mucus has been labelled with ^{35}S and the secretory output measured at 15-min intervals by a washout method (*Phipps* and *Richardson* 1976). In dogs, the mucosa of an upper tracheal segment (see above) has been viewed through a dissecting microscope after being dried and sprayed with powdered tantalum (an inert metal) until a thin uniform layer coats the surface (Fig. 27A). The tantalum layer prevents ciliary dispersion of secretion issuing from the submucosal gland ducts, and the resulting accumulations of mucus elevate the tantalum to form hillocks (Fig. 27B). Counts of the hillocks appearing in unit time and measurements of hillock diameter provide an index of the rate of secretion (Figs. 28, 29) (*Davis* et al. 1982b).

5.6.1 Effects of Pulmonary C Fibres on Secretion

We think it likely that stimulation of pulmonary C fibres will be found to increase airway secretion, but the evidence so far is inconclusive. In experiments in cats, *Phipps* and *Richardson* (1976) injected phenyldiguanide into the right atrium but observed no increase in the output of radio-labelled glycoprotein. Administration of histamine aerosol to the lower airways was also ineffective, but administration of ammonia vapour to the lower airways evoked cough and an increase in tracheal glycoprotein output. *Phipps* and *Richardson* concluded that although it was unlikely that pulmonary C fibres (J receptors) or intrapulmonary irritant receptors had any influence on airway secretion, other lower respiratory afferents ('cough receptors') were undoubtedly effective. In dogs, by contrast, injection of capsaicin into the right atrium to stimulate pulmonary C fibres increased tracheal submucosal gland secretion measured by the hillock method, but the relatively long latency of the response (average 9 s), which exceeded the pulmonary circulation time, made it impossible to attribute the reflex effect to pulmonary C fibres with absolute certainty (*Davis, Roberts, Coleridge* and *Coleridge,* unpublished observations).

5.6.2 Effects of Bronchial C Fibres on Secretion

Selective stimulation of bronchial C fibres by injection of small doses of capsaicin or bradykinin into the right bronchial artery not only evoked tracheal contraction, but also caused a rapid increase in tracheal secretion (*Davis* et al. 1982b) (Figs. 27–29). Control injections of solvent had no effect. The reflex increase in secretion was abolished by cooling the lower cervical vagus nerves to 0°C (Figs. 28B, 29) and was restored by rewarming the nerves (Fig. 28C); the response was also abolished by atropine.

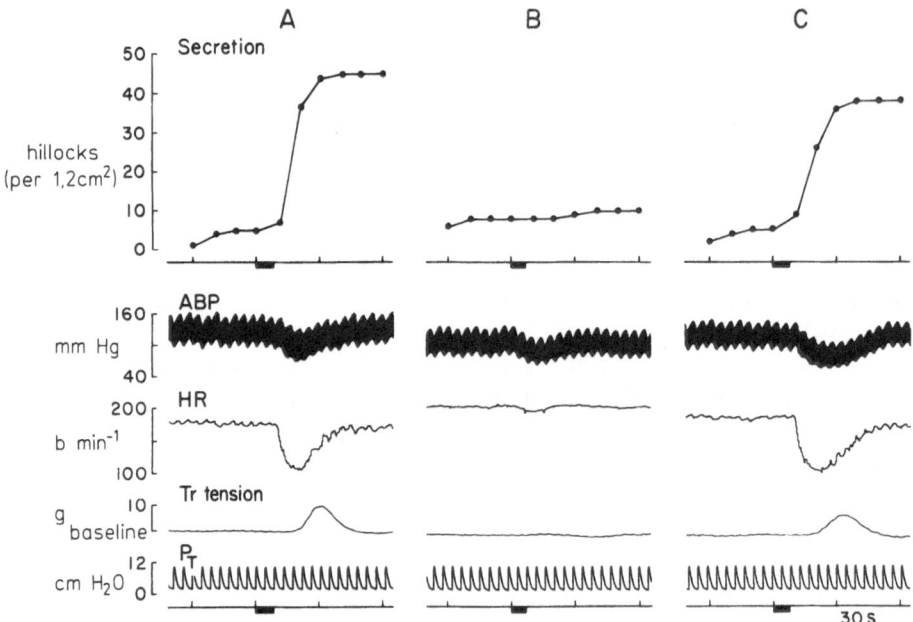

Fig. 28. Effects on tracheal submucosal gland secretion, tracheal tension, arterial blood pressure and heart rate evoked by injecting 3 μg capsaicin into the bronchial artery of a dog (at signal in *A*, *B* and *C*), and abolition of the effects by vagal cooling. Temperature of vagus nerves was: *A*, 37°C; *B*, 0°C; *C*, 37°C. Recurrent and para-recurrent nerves were already cut, so that cooling of mid-cervical vagus nerves had no effect on the segment's efferent nerve supply in the superior laryngeal nerves. *Secretion*, in hillocks per 1.2 cm^2 of mucosa; *ABP*, arterial blood pressure; *HR*, heart rate; *Tr tension*, tracheal tension in grams above a baseline set at 75 g; P_T, tracheal pressure. (*Davis* et al. 1982b)

Bronchial C fibres are the only pulmonary afferents to be stimulated by the small doses of chemical injected (*Kaufman* et al. 1980b; *Roberts* et al. 1981b; *Coleridge* et al. 1982a); hence there is good reason to conclude that they were responsible for the reflex increase in secretion. Since secretion usually coincided with tracheal contraction, it might be argued that the hillocks were not due to a sudden increase in secretion but merely represented previously secreted material that had been stored in the gland ducts until it was squeezed out by the contracting tracheal smooth muscle. However, hillocks sometimes appeared before tracheal contraction; indeed, in some experiments copious secretion occurred without any contraction at all – hence the response represented a true increase in the rate of secretion.

Sulphur dioxide administered to the lower airways of dogs, like ammonia vapour in cats (*Phipps* and *Richardson* 1976), causes cough and increased tracheal secretion, effects being abolished by cutting or cooling the lower cervical vagus nerves (*Hahn* et al. 1982b; *Roberts* et al. 1982b).

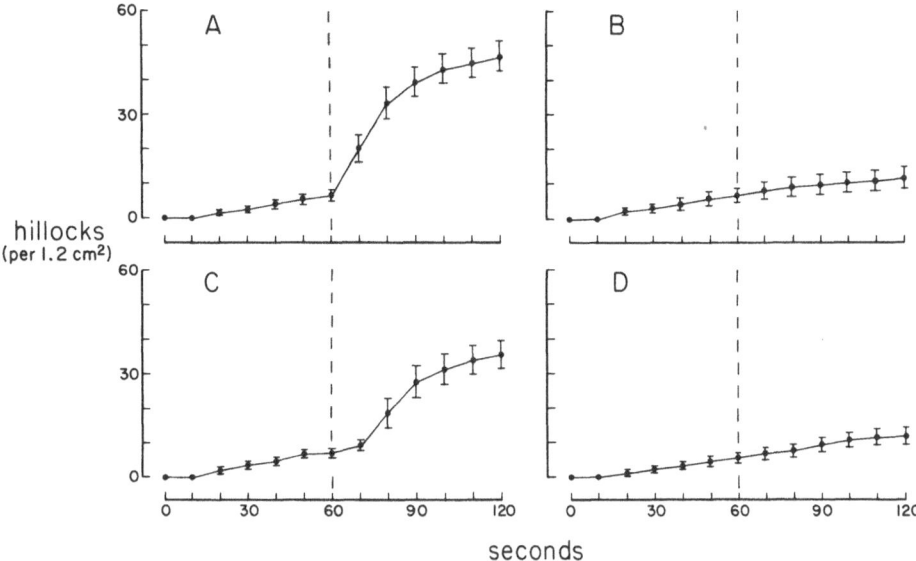

hillocks
(per 1.2 cm²)

Fig. 29A–D. Mean rate of appearance of fluid hillocks (per 1.2 cm² of mucosa) after injection of capsaicin or bradykinin into right bronchial artery, and the abolition of the response by cooling or cutting vagus nerves. **A, B** 3 μg capsaicin injected: **A** vagus nerves intact at 37°C (20 experiments); **B** vagus nerves cooled to 0°C (7 experiments). **C, D** 1.5 μg bradykinin injected: **C** vagus nerves intact at 37°C (14 experiments); **D** vagus nerves cut (2 experiments) or cooled to 0°C (5 experiments). *Vertical broken line* indicates time of injection. Values shown are means ± SE. (*Davis* et al. 1982b)

In parallel afferent fibre studies, pulmonary and bronchial C fibres were found to be the only vagal afferents to be stimulated by SO_2, the concentration required for threshold effects being 50 ppm. Concentrations of SO_2 from 100 to 500 ppm failed to stimulate rapidly adapting (irritant) receptors, and effects on slowly adapting pulmonary stretch receptors were, if anything, inhibitory (*Roberts* et al. 1982b). It seems not unreasonable to suggest that pulmonary as well as bronchial C fibres were involved in the secretory response.

5.7   Role of C Fibres in Cough and Irritant Sensations

The explosive expiratory act of coughing is perhaps the most dramatic reflex consequence of stimulating the afferent vagal nerves innervating the respiratory tract below the larynx; its function to expel secretions or inhaled irritants from the airways is immediately obvious. An increase in airway secretion usually accompanies coughing, and increases its effectiveness by acting as a vehicle in which foreign particles and chemicals can be transported up the airways for expulsion. Cough is also accompanied by bronchoconstriction (*Widdicombe* 1963, 1977a) and by irritant sensations transmitted by the vagus (*Noble* et al. 1970; *Guz* 1977b). The sensations can be crudely localized to the region from which the cough arises and may be described as having a tickling, scratching or burning quality, depending on the intensity of the stimulus. Coughing often gives temporary relief to such irritant sensations even when no secretions or inhaled materials appear to be expelled; in this sense 'tickling in the tubes' is transiently relieved by coughing in much the same way as itching of the skin is relieved by scratching. Repetitive, non-productive cough is often characteristic of lower airway diseases such as asthma.

In a review which, although brief, remains one of the most useful expositions of the cough reflex, *Bucher* (1958) suggested that there were no specific cough receptors as such in the mucous membrane of the lower respiratory tract, but that a sufficient increase in the general level of afferent stimulation or irritation by any factor could trigger cough. *Bucher* also proposed that the expiratory facilitation provided by input from slowly adapting pulmonary stretch receptors was essential for the induction of cough. This hypothesis has received some support from the results of recent experiments in rabbits, in which selective block of pulmonary stretch receptors by the brief administration of high concentrations of SO_2 abolished the cough reflex induced by application of chemical and mechanical stimuli to the lower airways and larynx (*Hanacek* et al. 1980; *Sant'Ambrogio* et al. 1980). Nevertheless, although there now appears to be some support for the notion that no single category of afferent is solely responsible, there is good evidence that certain categories of afferent are critically involved in the induction of coughing from particular parts of the lower airways. Thus, experiments on animals indicate that cough triggered by particulate irritants in the lower respiratory tract arises predominantly from stimulation of rapidly adapting receptors that are present in large numbers in the region of the carina: these are, indeed, the only afferents to which the specific appellation 'cough receptor' has been applied (*Fillenz* and *Widdicombe* 1972; *Widdicombe* 1954a, 1977a). The major reflexogenic area for cough triggered by chemical irritants such as SO_2 and ammonia is located more peripherally, in the intrapulmonary airways (*Widdicombe* 1954a, 1963, 1964).

Present evidence suggests that C fibres supply the critical afferent input when cough and sensations are evoked by chemical stimulation within the lung. The evidence, some of which is necessarily indirect, is based on the results of reflex studies in human subjects and experimental animals and on their interpretation in the light of vagal action potential studies in animals. Although cough has never been described as part of the pulmonary chemoreflex in animals, it appears to be a prominent component of the pulmonary chemoreflex response that is induced in man by intravenous or pulmonary arterial injection of lobeline (*Eckenhoff* and *Comroe* 1951; *Bevan* and *Murray* 1963; *Stern* et al. 1966; *Jain* et al. 1972). Thus *Eckenhoff* and *Comroe* (1951) described cough as the most prominent respiratory response to injection of lobeline. In a later study by *Jain* et al. (1972), in which a smaller dose of lobeline was injected, cough occurred only after several seconds of apnoea, and still smaller doses of lobeline evoked apnoea only. Retrosternal burning sensations, not sufficiently severe to be alarming, were described by the subjects in both studies; some subjects also described sensations referred to the lower throat and nose (*Jain* et al. 1972). The cough and retrosternal burning sensations reported by *Eckenhoff* and *Comroe* (1951) occurred as the lobeline was passing through the pulmonary circulation, 5–10 s before the onset of the typical increase in rate and depth of breathing evoked by stimulation of arterial chemoreceptors when lobeline gained the systemic circulation. Brady-cardia has also been reported as a component of the short-latency response to injection of lobeline in man (*Bevan* and *Murray* 1963). Both *Dawes* and *Comroe* (1954) and *Jain* et al. (1972) attributed the cough and respiratory sensations to stimulation of the pulmonary nerve endings responsible for initiating the pulmonary chemoreflex. Although others have suggested that the afferent endings responsible for these effects are located in the pulmonary artery rather than in the lung itself (*Bevan* and *Murray* 1963; *Stern* et al. 1966), there is no convincing electroneurographic evidence to support this hypothesis, and *Paintal*'s (1973) suggestion that the immediate effects of injecting lobeline into the right heart are adequately explained by the stimulation of C fibres in the lung seems the most reasonable one at present.

Inhalation of bradykinin aerosols also evokes coughing, or an urge to cough, and sensations of irritation, tickling and rawness in human subjects (*Lecomte* et al. 1962; *Simonsson* et al. 1973); the sensation of rawness in the airways suggested that bradykinin aerosol had passed through the larynx to the lower respiratory tract. There is experimental support for the notion that airway C fibres are involved in this response in man: thus, administration of bradykinin aerosols to dogs in concentrations similar to those used in the human studies vigorously stimulates C fibres in the extra- and intrapulmonary airways but has no significant effect on other vagal afferents in the lower respiratory tract (*Coleridge* et al. 1983).

Marked coughing has also been evoked in conscious dogs by adminis-tration of SO_2 to the lower airways through a tracheostomy (*Hahn* et al. 1982b). Administration of similar concentrations of SO_2 to the lower air-ways in anaesthetized dogs stimulated both bronchial and pulmonary C fibres, and also C fibres in the extrapulmonary airways, but had little effect on irritant receptors (*Roberts* et al. 1982b). Coughing evoked in conscious dogs by prolonged inhalation of SO_2 followed a characteristic time course, episodes of rapid shallow breathing occurring at intervals of 2–3 min, each acceleration of breathing culminating in a brief bout of coughing (*Hahn* et al. 1982b). Action potential studies revealed that the stimulation of afferent vagal C fibres from the lungs and lower airways during prolonged inhalation of SO_2 followed a similarly episodic time course (*Roberts* et al. 1982b). In conscious dogs both coughing and the episodic tachypnoea (together with the other respiratory and circulatory effects described in earlier sections) are abolished by cooling the vagus nerves to 0°C. There can be little doubt that afferent vagal C fibres arising from the lower respiratory tract provided the critical input for the coughing in the above-mentioned studies.

5.8 Reflex Cardiovascular Depressor Effects

Vagal afferent input from the lower respiratory tract and lungs has been known since the middle of the nineteenth century to have an inhibitory effect upon the heart (see *Brodie* and *Russell* 1900); indeed, *Brodie* and *Russell* claimed that of all the afferent vagal inputs capable of evoking reflex cardiac slowing, that from the lungs was most potent. They found that central stimulation of even a single small hilar branch of the pulmonary vagus in dogs evoked an immediate bradycardia; a similar bradycardia of rapid onset was evoked by administration of high concentrations of irritant gases to the lower airways and could no longer be obtained after the pulmonary branches of the vagus were cut. *Brodie* and *Russell* suggested that the lung afferents responsible for this reflex cardiac inhibition were identical with those initiating the apnoea, bradycardia and peripheral vasodilatation evoked in cats by intravenous injection of serum or egg-white (*Brodie* 1900). What was later to be called the 'pulmonary depressor chemoreflex' (*Dawes* and *Comroe* 1954) has since been demonstrated in several species, and it is now clear that pulmonary C fibres are the only afferents whose latency of response is sufficiently short to account for the rapid onset of these powerful cardiovascular reflexes. Bronchial C fibres also exert depressor effects, although their reflex influence on the heart and circulation appears to be less powerful than that of pulmonary C fibres.

5.8.1 Role of Pulmonary C Fibres

5.8.1.1 Cardiac Effects
The cardiac effects of stimulating pulmonary C fibres are seen in their most striking form in the pulmonary chemoreflex. The bradycardia resulting from right atrial injection of capsaicin or phenyldiguanide is

immediate and profound, and undoubtedly accounts for the major part of the dramatic fall in arterial blood pressure (Figs. 5, 16). Because both afferent and efferent limbs of the cardiac reflex travel in the cervical vagus nerves, the conventional techniques of cutting or cooling the vagi at this level do not necessarily help to identify the afferent pathways. Nevertheless, in cats the bradycardia induced by phenyldiguanide is not abolished until the vagus nerves are cooled to below 3°C (*Dawes* et al. 1951), indicating that in this species C fibres contribute to both afferent and efferent limbs of the reflex. In dogs, however, the reflex cardiac effects of capsaicin are said to be abolished at a vagal temperature of 8°–10°C, a finding that was explained by the observation that conduction in cardiac efferents in dogs appeared to be blocked at this temperature (*Porszasz* et al. 1957).

A major component of the cardiac chemoreflex in cats is a fall in right ventricular output that cannot be explained by a reduction in venous return or an increase in pulmonary vascular resistance (*Barer* and *Nusser* 1958); it appears to be a primary vagal effect on the heart. Vagal inhibition begins within 2–3 s of right atrial injection of capsaicin and often produces a period of asystole lasting for several seconds (Fig. 16) (*Coleridge* et al. 1964b, 1965, 1968); restoration of sinus node activity is often accompanied by several cycles of atrioventricular block (*Brender* and *Webb-Peploe* 1969). This vagal inhibition may be more pronounced in spontaneously breathing animals, when it is accompanied by arrest of breathing, than in artificially ventilated animals, when phasic lung inflation is uninterrupted (*Coleridge* and *Coleridge,* unpublished observations). This difference is probably accounted for by the difference in pulmonary stretch receptor activity in the two situations. Slowly adapting pulmonary stretch receptors promote tachycardia (*Daly* 1972); during apnoea their influence is withdrawn, whereas during artificial ventilation their rhythmic input may to some extent oppose the bradycardia evoked by chemical stimulation of pulmonary C fibres.

Several investigators have suggested that baroreceptors located in the extrapulmonary parts of the pulmonary artery play a major part in the cardio-inhibitory response to capsaicin (*Porszasz* et al. 1957; *Bevan* 1962; *Brender* and *Webb-Peploe* 1969). Any such effect is probably trivial, however, and although sensitization of pulmonary arterial baroreceptors could conceivably contribute to the inhibitory response once it is under way, any increase in baroreceptor activity is minor and of late onset (*Coleridge* et al. 1964b), and only pulmonary C fibres with their short latency of response can account for the immediate effects (see *Coleridge* and *Coleridge* 1979).

Vagal inhibition of the heart is also evoked when pulmonary C fibres are stimulated by large lung inflations – an effect that is now the focus of a good deal of attention because of its clinical relevance in certain types of lung injury and disease, in which positive end-expiratory pressure is employed to improve gas exchange. Under these conditions reflex vagal cardiac inhibition is an added complication to the mechanical embarrassment of cardiac function by ventilation at high intrathoracic pressures (*Cassidy* et al. 1978, 1979; *Cassidy* and *Mitchell* 1981). Since the time of Hering, lung inflation has been known to cause complex changes in heart rate, and *Brodie* and *Russell* (1900), in referring to unpublished studies of the cardiac effects of lung inflation, drew attention to Hering's observation in 1871 that if the air breathed by an animal was delivered to the airway at a small positive pressure, heart rate increased, but if it was delivered at a higher pressure, heart rate decreased. Some 60 years later, *Anrep* et al. (1936) confirmed that the cardiac effects could be reversed, tachycardia being converted to bradycardia as the lung was progressively inflated. The decreased heart rate evoked by large lung inflations is accompanied by a depression of cardiac contractility (*Glick* et al. 1969; *Greenwood* et al. 1977; *Cassidy* et al. 1979) which probably results from withdrawal of sympathetic efferent output (*Greenwood* et al. 1977).

These cardiac effects appear to be another example of a reflex response to inflation that undergoes a reversal of sign when the influence of low-threshold slowly adapting pulmonary stretch receptors is overridden by that of the higher threshold pulmonary C fibres. The inflation threshold for this reversal is of considerable interest. In studies in dogs, the stimulus was equated with inflation or airway pressure regardless of whether the chest was open or intact or whether one or both lungs were inflated (*Glick* et al. 1969; *Hainsworth* 1974; *Greenwood* et al. 1977; *Cassidy* et al. 1979). Results in some cases, therefore, may have given rise to the quite false impression that unphysiologically high inflation pressures are required for the elicitation of reflex depressor effects on the heart. For example, in spontaneously breathing dogs intubated by a tracheal divider that allowed independent inflation of each lung, *Hainsworth* (1974) found that during progressive lung inflation a pressure of 30–40 cm H_2O was required to convert tachycardia to bradycardia. It seems likely that under these conditions the distended lung was inflated by no more than two or three times the normal tidal volume, an increase in volume that agrees well with what is known of the inflation threshold for pulmonary C fibre stimulation (*Coleridge* et al. 1965, 1968; *Armstrong* and *Luck* 1974; *Coleridge* and *Coleridge* 1977b; *Kaufman* et al. 1982a). When an inflation pressure of 40 cm H_2O was applied to one lung in dogs with intact chest, the average reduction in heart rate was 89 beats min^{-1} (*Hainsworth* 1974). In dogs with open chest the inflation pressure threshold for a decrease in heart rate, and in dP/dt_{max} in the paced heart, was 13–20 cm H_2O (*Greenwood* et al. 1977). In both these studies the afferent pathway for the reflex clearly originated in the lung, for effects were abolished or greatly diminished after pulmonary innervation was damaged by three breaths of steam.

Complications arising from changes in blood gas tension and venous return have been avoided in some studies by distending the non-perfused lung or lungs. In experiments in open-chest dogs on cardiopulmonary by-pass, *Glick* et al. (1969) distended the lungs to a pressure of 20 mm Hg and found that heart rate decreased on average by 22% and that the pressure developed by the isovolumetrically contracting left ventricle decreased by 14%, the pressure threshold for cardiac inhibition being 10 mm Hg. To determine whether cardio-inhibitory effects were persistent, *Cassidy* et al. (1979) distended the vascularly isolated left lung in open-chest dogs for 15 min or more. Heart rate and stroke volume had decreased by 24% and 20% respectively 15 s after the onset of an inflation to 30 cm H_2O; heart rate returned to control level after 1 min, but stroke volume remained depressed throughout the 15-min period.

5.8.1.2 Effects on Peripheral Resistance

Brodie (1900) observed that the fall in arterial pressure induced by intravenous injection of serum or egg-white in cats was still present, although much smaller, after cardiac effects were prevented by atropine; moreover, the decrease in pressure was accompanied by an increase in the volume of a segment of intestine. *Brodie* therefore suggested that peripheral vaso-dilatation was a component of the chemoreflex from the lungs. Similarly, *Coleridge* et al. (1964b) found that the hypotension induced by capsaicin in dogs was not totally abolished by atropine, and suggested that pulmonary C fibres evoked a reflex peripheral vasodilatation. These speculations were confirmed by *Brender* and *Webb-Peploe* (1969), who observed that right atrial injection of capsaicin in dogs evoked a vagally mediated reflex dilatation of hind-limb resistance vessels and splenic capacity vessels, with an onset that often coincided with the onset of reflex bradycardia.

It seems likely that pulmonary C fibres contribute to the peripheral vasodilatation evoked by large lung inflations in dogs whose systemic circulation is artificially perfused to keep systemic arterial pressure and blood gas tensions constant. An overall decrease in systemic vascular resistance, with dilatation of hind limb resistance vessels, has been a general finding, the effects being abolished by cervical vagotomy (*Salisbury* et al. 1959; *Daly* et al. 1967; *Daly* and *Robinson* 1968; *Glick* et al. 1969; *Lloyd* 1978) or by denervation of the lungs at the hilum (*Daly* et al. 1967). *Daly* and *Robinson* (1968) showed that the reflex vasodilatation also involved the cutaneous and splanchnic circulations. Involvement of afferent vagal C fibre input from the lungs was suggested by the observation that large inflations often caused reflex vasodilatation when the vagus nerves were cooled to between 2° and 7°C (*Daly* et al. 1967). However, reflex vasodilator effects that first become apparent at lung volumes below FRC (*Daly* et al. 1967; *Daly* and *Robinson* 1968; *Lloyd* 1978), and hence below those likely to cause recruitment of pulmonary C fibres, are clearly due to stimulation of slowly adapting pulmonary stretch receptors alone. It seems reasonable to conclude that progressive inflation of the lung evokes progressive reflex vasodilatation: at small lung volumes effects are due entirely to input from slowly adapting pulmonary stretch receptors; at larger volumes, the additional input from pulmonary C fibres contributes to the reflex vasodilatation. In the presence of normal baroreceptor reflexes, the vasodilatation evoked by even large lung inflations appears to adapt rapidly when the lungs are inflated by a steady pressure (*Glick* et al. 1969). However, when the lungs are inflated phasically to reproduce the distortion provided by positive pressure ventilation, reflex vasodilatation is well maintained and is linearly related to peak inflation pressures, 6—15 cm H_2O representing the threshold of the response (*Lloyd* 1978).

Of the reflex effects of lung inflation considered so far, those on breathing, airway smooth muscle and heart rate appear to conform to the general rule that changes evoked by stimulation of low-threshold pulmonary stretch receptors are reversed by the recruitment of what were once called 'high threshold inflation receptors', i.e. pulmonary C fibres. The effects of inflation on the vasomotor centre, however, do not conform to this rule, and instead provide an instance of a reflex in which the influence of these two very different afferent inputs is synergistic, rather than antagonistic.

5.8.2 Role of Bronchial C Fibres

5.8.2.1 Cardiac Effects

Stimulation of bronchial C fibres by injection of bradykinin or capsaicin into a bronchial artery in artificially ventilated dogs evokes cardiac slowing in about 75% of experiments (Figs. 24, 28) (*Roberts* et al. 1981b; *Coleridge* et al. 1982a; *Davis* et al. 1982b); hence bronchial C fibres appear to have cardiac effects similar to those of pulmonary C fibres. However, if the cardiac effects of bronchial and pulmonary C fibres are considered relative to their effects on airway smooth muscle, bronchial C fibres are seen to have less effect on heart rate. Thus, in paired experiments, stimulation of bronchial C fibres by capsaicin evoked on average a 1.2% reduction in heart rate for each 1 g increase in tracheal tension, whereas stimulation of pulmonary C fibres by capsaicin evoked on average a 4.4% reduction in heart rate for each 1 g increase in tension; this differential effect was highly significant (*Coleridge* et al. 1982a). Hence bronchial C fibres appear to have weaker cardiodepressor effects and stronger airway smooth muscle effects than pulmonary C fibres.

In spontaneously breathing dogs, the cardio-inhibitory effects of bronchial C fibres were less easy to demonstrate; heart rate decreased in only a third of experiments and increased in the remainder (*Coleridge* et al. 1983). Cardio-acceleration may have been secondary to the dominant influence of pulmonary stretch receptors stimulated by tachypnoea, but most probably it resulted from the interplay of several afferent inputs set in train by the effects of bradykinin. Whatever the explanation, however, it seems clear that any primary inhibitory influence of bronchial C fibres on heart rate is readily overcome by secondary reflexes having cardio-acceleratory effects. The reflex influence of bronchial C fibres on cardiac contractility has not yet been examined.

5.8.2.2 *Effects on Peripheral Resistance*
So far there is no convincing evidence that the stimulation of bronchial C fibres evokes peripheral vasodilatation, or indeed has any effect on peripheral vascular resistance. Stimulation of bronchial C fibres in the intra- and extrapulmonary airways by bronchial arterial injection or infusion of bradykinin or capsaicin, or by administration of bradykinin aerosol to the lower respiratory tract, produces conspicuous contraction of tracheal smooth muscle, increased secretion by tracheal submucosal glands and, in some cases, cardiac slowing, often without any change in arterial blood pressure (Figs. 19, 21, 24C) (*Roberts* et al. 1981b; *Coleridge* et al. 1982a; *Davis* et al. 1982b). Bradykinin is a powerful vasodilator (*Garcia Leme* 1978) and the small reduction in arterial blood pressure observed in some experiments in which bradykinin was injected or infused into a bronchial artery probably resulted from the small amounts of the chemical that escaped degradation in the bronchial vascular bed and acted directly on peripheral vascular smooth muscle.

5.9 Effects on Somatic Motor Function: the J Reflex

The pulmonary chemoreflex response includes, as one of its most intriguing aspects, a profound depression of spinal reflex arcs. This depression was first described by *Ginzel* and his colleagues, who discovered by chance that intravenous injection of nicotine in cats depressed monosynaptic spinal reflexes, and that intravenous nicotine or phenyldiguanide caused a transient disappearance of decerebrate rigidity (*Ginzel* and *Eldred* 1969; 1977; *Ginzel* et al. 1969). The effects were peripheral rather than central in origin and appeared to be due to stimulation of vagal afferents in the cardiopulmonary region. It has been known for many years that a sudden increase in activity in visceral afferent pathways, including vagal afferents (*Schweitzer* and *Wright* 1937), carotid sinus baroreceptors (*Schulte* et al.

1959) and splanchnic and pelvic afferents (*Evans* and *McPherson* 1958), inhibits the monosynaptic and polysynaptic spinal reflexes. In recent years, however, the depression of spinal reflexes associated with the pulmonary chemoreflex response has commanded most attention.

Paintal has called this effect the 'J reflex' because the rapid onset of spinal reflex depression after right atrial injection of phenyldiguanide in cats (*Deshpande* and *Devanandan* 1970; *Paintal* 1970) indicates that it is triggered by activity in J receptors. Spinal reflexes are also depressed when phenyldiguanide is injected into the left atrium, beyond the pulmonary vascular bed (*Anand* and *Paintal* 1980; *Rosenthal, Coleridge* and *Coleridge,* unpublished observations), and undoubtedly vagal afferent nerve endings in the heart and further afield contribute to the J reflex once it is under way. Even so, depression of the knee jerk after right atrial injection of phenyldiguanide certainly outlasts any increase in afferent activity; it lasts for 30–60 s even when participation of cardiac afferents is ruled out by intrapericardial injection of local anaesthetics, and its time course closely parallels that of the respiratory effects (*Anand* and *Paintal* 1980). The accompanying apnoea and hypotension of the pulmonary chemoreflex do not contribute to the spinal reflex depression, since the J reflex is unaffected when apnoea and the consequent changes in blood gas tensions are avoided by artificial ventilation, and when hypotension is largely or completely prevented by administration of atropine (*Deshpande* and *Devanandan* 1970; *Ginzel* and *Eldred* 1969, 1977).

Central pathways for the J reflex ascend to the caudate nucleus and cingulate gyrus, and ablation of these areas abolishes the response (*Kalia* 1973). In the spinal cord, inhibitory pathways descend adjacent to the central canal and involve interneurones in the ventral pericanalicular region of the upper lumbar segment (*Ahluwalia* et al. 1977; *Rao* and *Devanandan* 1977). As might be expected of a reflex with a long central nervous pathway involving many synapses, the spinal reflex effects of phenyldiguanide disappear readily when the level of anaesthesia is deepened, even though reflex circulatory and respiratory effects may be well preserved (*Ginzel* and *Eldred* 1977; *Anand* and *Paintal* 1980). The J reflex has been demonstrated in conscious cats as a brief loss of motor function, accompanied by arrest of breathing (*Kalia* et al. 1973).

In Paintal's view this depression of the spinal motor outflow to skeletal muscle is perhaps the most important reflex regulatory function of pulmonary C fibres (J receptors) and has its greatest physiological significance in exercise (*Paintal* 1969, 1970, 1973; *Anand* and *Paintal* 1980). Paintal envisages J receptors, stimulated by an increase in pulmonary capillary pressure in exercise, as the sensors commanding a feedback system that limits muscular performance and prevents over-exertion. He envisages this mechanism as being engaged to maximal effect when exercise is performed under adverse physiological conditions (e.g. at high altitude, when exercise may cause pulmonary oedema), but he believes that it is of physiological significance at all grades of exercise (*Paintal* 1970; *Anand* and *Paintal* 1980). *Paintal*'s view of the functional significance of the J reflex is not universally accepted, however, partly because it is only one of many manifestations of viscerosomatic inhibition that may be evoked from a variety of visceral inputs, usually by the application of stimuli outside the normal range. Moreover, the J reflex itself has so far been demonstrated only when J receptors have been activated by foreign chemicals such as nicotine and phenyldiguanide. In the case of phenyldiguanide, the doses injected into the right atrium of cats in a recent study by *Anand* and *Paintal* (1980) were held to produce an increase in firing in J receptors equivalent to that brought about by moderate exercise; in the event, even the smallest doses used to evoke the J reflex were sufficient to evoke an initial apnoea of about 10 s and, indeed, did not differ from the usual doses of phenyldiguanide ($10-40 \mu g\ kg^{-1}$) employed to evoke the pulmonary chemoreflex 'triad' in cats.

This association of spinal inhibition with arrest of breathing has been a topic of some discussion (*Koepchen* et al. 1977); effects on limb muscles comprise inhibition of α and γ motor neurones to both flexors and extensors (*Ginzel* et al. 1971), and effects on respiratory neurones and on respiratory muscles are of a similar wholesale and non-reciprocal character (*Schmidt* and *Wellhoner* 1970; *Koepchen* et al. 1977). Effects on respiratory neurones and on spinal motor neurones are thought likely to be the outcome of a single central phenomenon (*Koepchen* et al. 1977). Whether this phenomenon is 'all-or-none' in nature, or whether the motor components to limb muscles and to respiratory muscles can be affected selectively, and whether graded effects are possible in response to stimuli in the physiological range, is still unknown. Hence the role of the J reflex in exercise is still far from clear. Acceptance of *Paintal*'s view of the significance of the J reflex is hindered by the association, in both conscious and anaesthetized animals, of spinal inhibition and apnoea. Apnoea is the most unlikely, and certainly the most inappropriate, of any respiratory response to exercise, and the hypothesis that the J reflex has a physiological role in exercise would be more convincing if spinal reflex depression could be demonstrated in experimental circumstances in which tachypnoea was the only respiratory response.

The J reflex, and, indeed, the pulmonary chemoreflex as a whole, is regarded by some authorities as being of interest in mammals mainly as a curious evolutionary survival, possibly representing a 'sham-death' response whereby an animal may escape the attention of predators. This point of view is reinforced by the observation that this combination of reflexes is present in fish and thus seems to have developed relatively early in verte-

brate evolution, at a stage when there is little evidence for an active baroreceptor reflex (*Satchell* 1977; also see discussion of *Satchell*'s paper).
Thus phenyldiguanide, injected into the ductus cuvieri of the dogfish,
evokes the typical chemoreflex 'triad' of cessation of respiratory movements, bradycardia and hypotension; in addition, after a short interval,
the fish hangs limply from its harness for several seconds before resuming
swimming movements (*Satchell* 1977). Whether the somatic inhibition in
fish depends upon the integrity of afferent vagal C fibre pathways does
not seem to have been determined, but injection of phenyldiguanide was
found to evoke a burst of activity in afferent fibres in the branches of the
vagus nerve that supply the gills, and these were believed to be responsible
for the effects observed.

6 Functional Significance

Although non-myelinated fibres account for four-fifths of the afferent
vagal input from the lungs and lower airways, they are rarely included in
any comprehensive schema of the neural control of respiratory function.
Input from this small-fibre system is assumed to exercise no influence on
the control of breathing at rest (*Paintal* 1973) nor, indeed, under physiological circumstances in general (*Widdicombe* 1974a, b, 1977a, 1981;
Bradley 1977). Even in abnormal circumstances, input from the myelinated fibres of rapidly adapting (irritant) receptors is held to outweigh in
functional significance the contribution from the far more numerous C
fibres (*Widdicombe* 1974a, b, 1977a, 1981; *Nadel* 1980). *Paintal* (1969,
1970, 1973, 1977a, b) has repeatedly challenged the view that J receptors
are 'alveolar nociceptive endings' (*Widdicombe* 1974a, 1981) that exert
reflex actions only in pathological circumstances, but in his own view the
functional significance of J receptors is largely confined to the operation
of 'J reflex' (somatic inhibition and a sense of respiratory discomfort) as a
limiting factor in severe exercise (*Paintal* 1969, 1970, 1973; *Anand* and
Paintal 1980). Some reassessment of the functional significance of the
non-myelinated afferents is timely, since their distribution in the lower
respiratory tract is now recognized as being more widespread than was
formerly supposed and since much more is known of their afferent and
reflex properties.

6.1 Physiological Role

Without doubt arguments that afferent C fibres have no reflex function under normal circumstances have been influenced by the low and inconspicuous background discharge of these fibres, and by the relatively modest increases in discharge evoked by stimuli in the physiological range, as compared with the more massive increases induced by the customary bolus injections of foreign chemicals. However, the suggestion that a majority of J receptors are inactive in resting conditions (*Paintal* 1970, 1973) appears to be based on observations in artificially ventilated cats with chests widely open. When activity is recorded during spontaneous breathing, pulmonary C fibres in cats and dogs discharge several impulses in each respiratory cycle (average 1.9 impulses s^{-1}) and the resting discharge of bronchial C fibres, though less (0.8 impulses s^{-1}), is nonetheless appreciable (*Coleridge* and *Coleridge* 1977a, b). Such low frequency background discharges in vagal C fibres can be shown to exert appreciable reflex effects: thus, *Thoren* et al. (1975, 1977) found in cats that vagal afferents exert a considerable tonic depressor influence after sino-aortic denervation, the non-myelinated fibres accounting for 40%–80% of this effect.

When afferent C fibres are stimulated, even a low frequency of firing can evoke reflex responses. For example, results of electrical stimulation indicate that the afferent C fibres of the aortic nerve produce reflex depressor effects at much lower frequencies of stimulation than the A fibres (*Douglas* et al. 1956; *Kardon* et al. 1973). *Paintal* regards a frequency of 7 impulses s^{-1} as evidence of intense stimulation of J receptors, capable of evoking powerful reflex effects, and he suggests that a sustained input of less than 1 impulse s^{-1} can be of considerable reflex significance (*Paintal* 1970, 1973; *Anand* and *Paintal* 1980).

Experiments in both awake and anaesthetized animals provide strong evidence that afferent vagal C fibres from the lower respiratory tract have a role in physiological circumstances and that both the background (resting) discharge and the activity engendered by stimuli within the physiological range play a significant part in the neural control of respiratory function. There is some evidence, which we have already reviewed and which we do no more than mention here, that tonic activity in afferent C fibres from the lower respiratory tract contributes to the maintenance of vagal bronchoconstrictor tone (*Jammes* and *Mei* 1979; *Roberts* et al. 1981b, 1982a). But perhaps the most striking physiological function of these afferents, and one for which evidence has been gradually accumulating over the last 10–15 years, is their influence on breathing rate, in particular their ability to shorten T_E as well as T_I and thus to promote a pattern of breathing that is relatively rapid and shallow.

6.1.1 Influence of Resting Discharge on Breathing Rate

Discussion of the role of vagal afferents in determining the rate of breathing under normal circumstances is usually confined to 'volume feedback' from the slowly adapting stretch receptors with myelinated fibres. The discharge of these fibres at a critical level of inspiration provides an 'inspiratory off-switch' that determines T_I, and their discharge at FRC lengthens the expiratory pause and determines T_E (*Clark* and *von Euler* 1972; *Bartoli* et al. 1973; *Bradley* 1977; *Trenchard* 1977). Any influence of afferent C fibres on the rate of breathing at rest is held to be unlikely.

A hypothesis that breathing rate at rest is determined by the balance between an inhibitory input from the lower respiratory tract, blocked at a vagal temperature of 8°C, and an excitatory input blocked only at 2°–3°C, was developed by *Hammouda* and *Wilson* (1935a, b, 1939) in the course of studies of the effects of vagal cooling on the pattern of breathing in rabbits and dogs. This hypothesis has been largely neglected, although more recent evidence to support it and to confirm the original observations is not hard to find. In rabbits and dogs, for example, cooling the vagus nerves to 7°–8°C is reported to increase minute ventilation and to decrease end-tidal CO_2, both values reverting to control levels when the nerves are cooled further or cut (*Karczewski* and *Widdicombe* 1969a; *Phillipson* et al. 1973).

Hammouda and *Wilson* believed that the small-fibre pathway exerted an excitatory influence only on breathing rate. In the first of their studies they used Head's method and allowed the vagus nerves of rabbits to warm gradually after being packed in ice. They observed an increase in breathing rate several minutes before the return of the Hering-Breuer inflation reflex. Developing methods for controlling nerve temperature more accurately, they found that vagal cooling to 8°–5°C caused acceleration of breathing in dogs, and breathing did not become slow and deep until the temperature of the nerves was reduced to below 5°C (*Hammouda* and *Wilson* 1939). The effects were attributed to an acceleratory influence of small-fibre endings in the lungs, because acceleration of breathing was invariably evoked when the pulmonary branches of the vagus were stimulated during partial cooling or compression of the nerve trunk, whereas stimulation of the abdominal vagus was without effect. The observations of *Hammouda* and *Wilson* concerning the effects of vagal cooling on breathing rate in dogs, though forgotten, were confirmed some 40 years later in experiments in conscious dogs with exteriorized vagus nerves. Again, an acceleration of breathing was observed at vagal temperatures of 8°–5°C, the pattern of breathing at this range of temperatures being described as rapid and shallow, and giving way to slow deep breathing upon further cooling (*Fishman* et al. 1973; *Phillipson* et al. 1973).

The acceleration of breathing at low vagal temperatures was attributed to tonic activity in rapidly adapting (irritant) receptors, on the mistaken assumption that the fibres of these receptors were smaller and more resistant to cooling than those of the slowly adapting stretch receptors, believed to comprise the inhibitory pathway. It is now clear, however, from conduction velocity measurements of large numbers of afferent fibres in dogs, that the fibres supplying rapidly adapting receptors are myelinated (*Sampson* 1977); hence their input is likely to be blocked completely over the temperature range at which acceleration of breathing was observed. It seems likely, as *Hammouda* and *Wilson* (1939) suggested, that a specific fine-fibre pathway supplies the acceleratory influence, and that C fibre input from the lower respiratory tract makes the major contribution. One may postulate that the excitatory pathway is more susceptible than the inhibitory pathway to the blunting effects of anaesthesia, a factor that might explain the observation that anaesthesia potentiates Hering-Breuer reflex inhibition (*Bouverot* et al. 1970; *Phillipson* et al. 1971).

When the vagus nerves are cooled gradually in rabbits and cats an intermediate phase of increased breathing frequency is not apparent, and breathing becomes progressively slower and deeper (*Karczewski* and *Widdicombe* 1969a; *Miserocchi* et al. 1978). However, when the vagus nerves are cut after being first cooled to 6°C to block conduction in myelinated fibres, T_E becomes markedly prolonged in rabbits and prolonged to a lesser degree in cats — effects attributed to interruption of lower airway C fibres (*Miserocchi* et al. 1978). Nevertheless, a tonic excitatory influence of lower airway C fibres cannot always be demonstrated in anaesthetized animals, and other investigators have concluded that, in the absence of pathological conditions of the lung, block of myelinated fibres in the rabbit vagus does not reveal any tachypnoeic influence of non-myelinated ones (*Guz* and *Trenchard* 1971; *Trenchard* et al. 1972).

6.1.2 Afferent C Fibres and the Tachypnoea of the CO_2 Response

Addition of CO_2 to the inspired air is not, strictly speaken, a 'physiological' stimulus to breathing, yet it is often used to demonstrate physiological mechanisms of respiratory control. Studies in man, as well as in experimental animals, indicate that the ventilatory response to CO_2 is markedly depressed after interruption of the vagus; T_E no longer shortens, and the increase in ventilation comes to depend almost entirely on the increase in V_T resulting from stimulation of the central chemoreceptors (*Guz* et al. 1966; *Richardson* and *Widdicombe* 1969; *Phillipson* et al. 1970; *Bradley* et al. 1974). The tachypnoea of the ventilatory response to CO_2 was formerly attributed to volume feedback from slowly adapting stretch receptors, but there is now evidence that this tachypnoea, together with

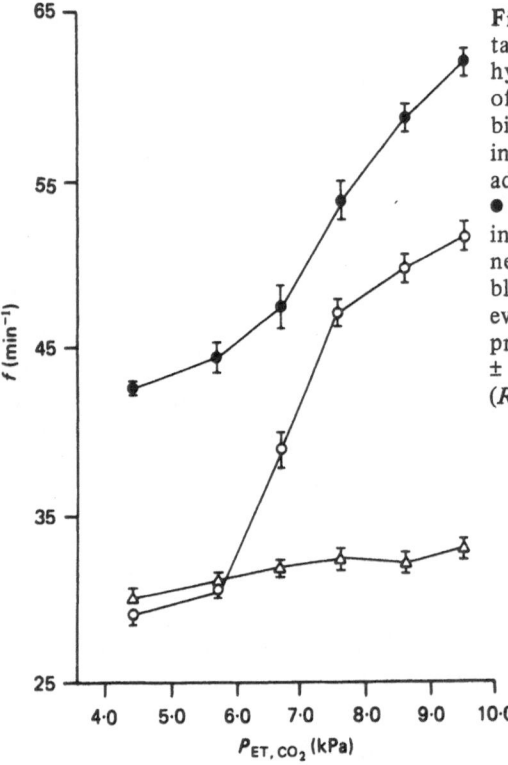

Fig. 30. Role of pulmonary C fibres in tachypnoeic response to CO_2: effect of hypercapnia on breathing frequency *(f)* of a pentobarbitone-anaesthetized rabbit. Hypercapnia produced by rebreathing from a closed system to which O_2 added to keep PaO_2 above 250 mm Hg. ● Left vagus nerve cut, right vagus nerve intact. ○ Left vagus nerve cut, right vagus nerve differentially blocked (anodal block) so that only the C wave of the evoked compound action potential was present. △ Both vagus nerves cut. Mean ± SEM of five re-breathing periods. (*Raybould* and *Russell* 1982)

the shortening of T_E which is its characteristic feature, persists when conduction is blocked in myelinated vagal fibres (*Phillipson* et al. 1973; *Raybauld* and *Russell* 1982). In anaesthetized rabbits, for example, breathing becomes slower and deeper when conduction in myelinated fibres is blocked by anodal polarization; nevertheless, breathing rate increases during CO_2 breathing, with a slope not different from that obtained when conduction in myelinated fibres is intact (Fig. 30; *Raybould* and *Russell* 1982). Moreover, in conscious (*Phillipson* et al. 1973) and anaesthetized (*Cross* et al. 1976) dogs and in conscious and anaesthetized human subjects (*Cross* et al. 1976) the increase in breathing rate and the decrease in T_E in response to inhaled CO_2 become exaggerated as the strength of the Hering-Breuer inflation reflex decreases, providing persuasive evidence that the input of slowly adapting stretch receptors acts to some extent to oppose, rather than promote, the breathing rate response to CO_2. The tachypnoea of the CO_2 response in dogs is at a maximum at a vagal temperature of 8°–4°C (*Phillipson* et al. 1973). Thus, the weight of evidence indicates that the tachypnoea of the CO_2 response is a function of input in afferent vagal C fibres, and *Raybould* and *Russell* (1982) suggest that the input arises from the lung. The proven ability of lower respiratory tract C fibres to induce reflex tachypnoea and to decrease T_E is certainly compatible with such a role.

The mechanism by which CO_2 administration increases activity in these C fibres is uncertain. Stimulation may be secondary to the increase in V_T or may be due to a specific sensitivity to CO_2. Large inflations of the lung stimulate pulmonary C fibres and even larger ones stimulate airway C fibres (*Coleridge* et al. 1965, 1968; *Coleridge* and *Coleridge* 1977b; *Kaufman* et al. 1982a); hence the tachypnoea of the CO_2 response may be secondary to stimulation of lung C fibres by increased V_T. So far, however, the sensitivity of afferent C fibres to large lung inflations has been examined only during positive pressure inflation. It remains to be seen whether afferent C fibres from the lungs and airways are stimulated when tidal volume increases during spontaneous breathing, and whether volume feedback is indeed a major factor in the CO_2 effects.

Evidence for a specific CO_2 sensing mechanism in the lungs of dogs has been obtained in experiments in which the pulmonary and systemic circulations were separately perfused and the chest was closed to allow spontaneous breathing (*Sheldon* and *Green* 1982). These investigators found that increasing pulmonary arterial PCO_2 from 35 to 80 mm Hg caused a vagally mediated increase in V_E. The effect was later found to be augmented by increasing pulmonary blood flow at high pulmonary arterial PCO_2, and according to the authors' estimate the total increase in ventilation induced by this means could account for 60% of the hyperpnoea of exercise (*Green* and *Sheldon* 1983). Since systemic $PaCO_2$ was kept constant, any increase in lung afferent activity could not have been secondary to a change in the pattern of breathing initiated by central medullary chemoreceptors and appears to have been triggered by the action of CO_2 on afferent endings in the lung. Nevertheless, the sensitivity of lung C fibres to CO_2 is still controversial. Some authors have suggested that a majoritiy of 'bronchopulmonary C fibres' exhibit some degree of CO_2 sensitivity (*Delpierre* et al. 1981). So far, however, there is no convincing evidence from afferent studies that C fibres with endings in the lower respiratory tract have a CO_2 sensitivity that accounts for the tachypnoeic response to inhaled CO_2. In the study by *Delpierre* et al. (1981), for example, C fibres sensitive to the range of CO_2 encountered in studies of the ventilatory CO_2 response were in the minority; they adapted rapidly to a steady-state increase in CO_2, were stimulated equally by the off-transient (Fig. 12) and their origin in the respiratory tract was not confirmed. Obviously, the part played by afferent vagal C fibres in the ventilatory response to CO_2 remains a topic of considerable interest.

6.1.3 Role of Afferent C Fibres in Exercise

Pulmonary C fibres are probably susceptible to stimulation by a number of the physiological changes that occur during muscular exercise. Thus pulmonary C fibres appear to be sensitive to pulmonary circulatory changes (*Coleridge* and *Coleridge* 1977a; *Anand* and *Paintal* 1980) and to increases in tidal volume, though the latter effect has so far been demonstrated only in response to positive pressure inflation (*Coleridge* et al. 1965; *Armstrong* and *Luck* 1974; *Coleridge* and *Coleridge* 1977b; *Kaufman* et al. 1982a).

Paintal initially laid great stress on pulmonary congestion as a stimulus to J receptors in exercise, regarding as their most significant afferent function a sensitivity to increased alveolar interstitial tension resulting from increased pulmonary capillary pressure (*Paintal* 1969, 1970, 1973). Avoiding the question of whether pulmonary congestion is a necessary accompaniment of exercise, he confined his discussion to exercise of a severe degree or exercise undertaken at high altitude. A physiological role for pulmonary C fibres in exercise is more strongly supported by the observation in cats that impulse activity increased when blood flow to the appropriate lung lobe increased (*Anand* and *Paintal* 1980). A relationship between pulmonary C fibre discharge and the degree of filling of the pulmonary vascular bed is also suggested by a significantly higher rate of discharge during spontaneous breathing than during positive pressure ventilation when the chest is open (*Coleridge* and *Coleridge* 1977a). Pulmonary C fibres are undoubtedly sensitive to pulmonary congestion (Fig. 15), but a more systematic examination of their response to changes in blood flow and volume in the physiological range is required before the possibility of their sensitivity to the pulmonary vascular changes of exercise can be fully accepted.

Although *Dickinson* and *Paintal* (1970) suggested that J receptors in cats might function as sensors of mixed-venous CO_2, *Paintal* did not pursue the topic of CO_2 sensitivity in a subsequent review (*Paintal* 1973). Instead he favoured the hypothesis that stimulation of J receptors in exercise was likely to be a result of purely circulatory changes. The possibility still remains that lung C fibres are stimulated by an increase in mixed-venous CO_2 in exercise but the evidence is far from conclusive.

What is the evidence that lower respiratory tract C fibres contribute to the tachypnoea of exercise? Even in the absence of vagal input, the rate of breathing increases in exercise, and in this respect the exercise response is quite unlike the ventilatory response to CO_2. Nevertheless, in experiments in conscious dogs performing treadmill exercise of varying degrees of severity, *Phillipson* et al. (1970) found that the rate of breathing was lower and V_T higher at each level of exercise after the vagus nerves were blocked by local anaesthetic. In a later brief report of the effects of differential vagal cooling during exercise, the permissive effect of vagal conduction on breathing rate was described as being present at low vagal temperatures, when only C fibres were conducting, and the investigators suggested that J receptors played a role in expercise tachypnoea (*Phillipson* et al. 1975a).

6.2 Role in Airway Defence Reflexes

The possibility that lower respiratory tract C fibres play some role in airway defence reflexes has been acknowledged, but until recently (*Coleridge* and *Coleridge* 1981) little emphasis has been placed on their contribution. The airway defence reflexes are essentially protective and their natural purpose appears to be to prevent or ameliorate the potentially harmful effects of foreign chemicals or particles that gain access to the respiratory tract with the inflow of air, a hazard that is not uncommon in everyday life. Hence they are 'physiological' in the sense of being one of the many composite patterns of reflex response that operate in emergencies, although they have a functional significance quite different from that of the continuously operating regulatory reflex mechanisms. When irritants are inhaled, reflexogenic sites in the nose and larynx act as a first line of defence to trigger the appropriate reflex responses, but once the irritant has reached the lower airways, further reflexes of a defensive nature are evoked. These include initial gasps or periods of apnoea, coughing, rapid shallow breathing, often cardiovascular depression, occasional sighs, bronchoconstriction, increased airway secretion and, in man, irritant sensations crudely located in a region beneath the sternum. Reflex studies in dogs now strongly support the hypothesis that pulmonary and bronchial C fibres, particularly the latter, play a major role in this composite reflex picture (*Russell* and *Lai-Fook* 1979; *Roberts* et al. 1981b, 1982b; *Davis* et al. 1982b; *Coleridge* et al. 1982a, 1983).

Various components of the airway defence reflexes can be recognized in disease of the lungs and airways, and often the endogenous release of autocoids that stimulate nerve endings appears to serve as the trigger. The responses evoked under these conditions can no longer be considered 'physiological', because the stimulus is part of a disease process, but the reflex picture does not differ. However, lung autocoids are often used simply as pharmacological tools to study defence reflexes originating in the lower airways; we discuss some of these studies in this section. We reserve the final section for a brief discussion of the functional significance of lower respiratory tract C fibres in lung disease.

6.2.1 Afferent C Fibres and Inhaled Irritants

In early studies powerful and noxious chemicals were administered to the lower airways, and only the immediate effects on breathing and on the cardiovascular system could be clearly distinguished as being of reflex origin (*Brodie* and *Russell* 1900; *Cromer* et al. 1933; *Whitteridge* 1948; *Banister* et al. 1950). For the most part these effects corresponded to what we now recognize as the reflexes evoked by stimulation of lower respiratory tract C fibres: apnoea, rapid shallow breathing and brady-cardia, which were abolished by vagotomy. In dogs, residual changes in breathing evoked after vagotomy could be attributed to input from an afferent pathway traversing the stellate ganglion (*Cromer* et al. 1933; *Banister* et al. 1950); such a pathway was also described in cats breathing high concentrations of SO_2 (*Widdicombe* 1954c). There is very little evidence so far that this alternative afferent pathway can evoke other components of the airway defence response, and the irritant sensations that are part of this response in man are known to be transmitted by the vagus nerves (*Noble* et al. 1970; *Guz* 1977b).

Commonly encountered airway pollutants such as SO_2 and cigarette smoke were formerly held to evoke defence reflexes from the lower re-spiratory tract mainly by stimulating irritant receptors with myelinated fibres (*Widdicombe* 1974a, b). Recent descriptions of airway defence re-flexes evoked in conscious and anaesthetized dogs by SO_2 (*Hahn* et al. 1982a, b; *Roberts* et al. 1982b) and by cigarette smoke (*Lee* et al. 1983) indicate major involvement of afferent C fibres, and in the case of SO_2 this has been confirmed by afferent studies (*Roberts* et al. 1982b). SO_2 delivered to the lower airways was found to evoke cough, rapid shallow breathing, and reflex increases in tracheal smooth muscle tone and sub-mucosal gland secretion (*Hahn* et al. 1982a, b; *Roberts* et al. 1982b). These reflex effects had an unusual time course, being marked at their onset, then subsiding and undergoing exacerbations at intervals of 2–3 min. In parallel afferent studies pulmonary and bronchial C fibres were stimulated by the same range of concentrations of SO_2 and with a latency of onset and a fluctuating time course similar to that observed for the re-flex effects. By contrast, irritant receptors were usually unaffected by the concentrations of SO_2 employed in these experiments, and those whose discharge increased showed only a minor and transitory response. Studies in cats (*Widdicombe* 1954a) and rabbits (*Davies* et al. 1978) also suggest that even high concentrations of SO_2 have relatively minor effects on ir-ritant receptors. Hence it seems likely that the reflex respiratory effects of SO_2 in the lower respiratory tract must be ascribed mainly to activity in afferent C fibres.

Respiratory effects evoked when cigarette smoke is delivered to the lower airways of dogs appear to be induced by stimulation of a wider variety of afferents, including irritant receptors and arterial chemorecep-tors, the latter being stimulated by the nicotine content of cigarette smoke (*Lee* et al. 1983). After the carotid bodies were denervated, ciga-rette smoke induced vagally mediated effects that appeared typical of afferent C fibre stimulation: namely, bradycardia, apnoea and rapid shal-low breathing. In some dogs large gasps or sighs occurred at the onset, suggesting involvement of irritant receptors also. Cigarette smoke consistently stimulates irritant receptors in rabbits (*Sellick* and *Widdicombe* 1971), but its effects on the corresponding afferents in dogs are rather less consistent, only about 20% of receptors being stimulated (*Sampson* and *Vidruk* 1975). This may explain why the prominent gasps, believed to be typical of the effects of irritant (rapidly adapting) receptor stimulation (*Knowlton* and *Larrabee* 1946; *Glogowska* et al. 1972), were not invariably present in the reflex experiments in dogs (*Lee* et al. 1983). In short, there is now good evidence that lower respiratory tract C fibres are responsible for some of the reflex effects of inhaled irritants previously attributed to irritant receptors with myelinated fibres.

6.2.2 Afferent C Fibres and Lung Autocoids

Histamine, either injected into the blood stream or administered as aerosol, has been widely used in experimental animals to evoke the rapid shallow breathing, occasional sighs or gasps and reflex contraction of airway smooth muscle that are typical of the airway defence response (*DeKock* et al. 1966; *Karczewski* and *Widdicombe* 1969c; *Fishman* et al. 1973; *Phillipson* et al. 1975a; *Bleecker* et al. 1976; *Winning* and *Widdicombe* 1976; *Miserocchi* et al. 1978). These observations, together with the finding that histamine stimulates irritant receptors with myelinated fibres (*Mills* et al. 1969, 1970; *Sellick* and *Widdicombe* 1971; *Sampson* and *Vidruk* 1975, 1978; *Sampson* 1977), were undoubtedly responsible for the general conviction that irritant receptors provide the major, if not the only, input for these and other reflex components of the airway defence response. The conviction was strengthened by the observation that in rabbits the reflex respiratory effects of histamine were virtually abolished by vagal cooling to 8°C (*Karczewski* and *Widdicombe* 1969c). In addition, J receptors (pulmonary C fibres), which were the only category of lower respiratory C fibre described at the time, were found to be insensitive to histamine (*Paintal* 1969, 1970, 1973), and histamine in its passage through the pulmonary circulation did not evoke the pulmonary chemoreflex. Indeed, when histamine was injected into a vein or the right atrium its ultimate site of action appeared to be in the conducting airways, since in both cats (*Winning* and *Widdicombe* 1976) and rabbits (*Karczewski* and *Widdicombe* 1969c; *Jain* et al. 1973) the respiratory effects began only after a latency of several seconds, and in rabbits were blocked or attenuated by administration of local anaesthetic aerosols, whereas the pulmonary chemoreflex evoked by phenyldiguanide was never impaired (*Jain* et al. 1973).

On the other hand, a number of observations suggested that the reflex respiratory effects of histamine could not be explained solely by activation of irritant receptors with myelinated fibres, and that afferent

C fibres were involved. Thus in dogs (*Fishman* et al. 1973; *Phillipson* et al. 1975a) and guinea-pigs (*Koller* and *Ferrer* 1973) the reflex respiratory effects of histamine did not disappear at a vagal temperature of 8°C; indeed, in conscious dogs the rapid shallow breathing evoked by histamine became more pronounced as the temperature was reduced from 8° to 4°C, and in both dogs and guinea-pigs ventilatory effects of histamine were not abolished until vagal temperatures reached 4°–0°C. Additional observations in dogs furnished supporting evidence that afferents other than irritant receptors contributed to the respiratory effects of histamine (*Dixon* et al. 1979a). Uncertainties about the afferent mechanisms responsible for the respiratory effects of histamine have been resolved by the observation that afferent vagal C fibres supplying the conducting airways are highly sensitive to histamine (*Coleridge* et al. 1978; *Coleridge* and *Coleridge* 1977b). These airway C fibres appear to be stimulated directly by chemicals, and their response to histamine is in no way dependent upon changes in lung mechanics.

Bradykinin has also been used to study the various components of the airway defence reflexes in man and experimental animals, and is another example of a lung autocoid to which bronchial C fibres are highly sensitive and pulmonary C fibres are not (*Kaufman* et al. 1980b; *Roberts* et al. 1981b). Moreover, irritant receptors in dogs are not stimulated by bradykinin (*Kaufman* et al. 1980b; *Roberts* et al. 1981b), and bradykinin lacks the direct bronchoconstrictor action of histamine on airway smooth muscle in dogs (*Waaler* 1961; *Garcia Leme* 1978). In healthy human subjects, and also in some asthmatics, inhalation of 0.01%–0.1% bradykinin aerosol induced an increase in airway resistance, which was prevented by atropine, and a sense of airway irritation and an urge to cough (*Simonsson* et al. 1973). The subjects inhaled bradykinin through a mouthpiece, so that a contribution of lower airway afferents to these effects cannot be established with certainty; nevertheless, the irritant airway sensations suggest that lower airway afferents were stimulated. In dogs injection of bradykinin into a bronchial artery or administration of 0.1% aerosol to the lower airways induced rapid, shallow breathing, increased airway smooth muscle tone and increased tracheal submucosal gland secretion by stimulating bronchial C fibres; cardiac slowing sometimes, though not invariably, accompanied these responses (*Roberts* et al. 1981b; *Davis* et al. 1982b; *Coleridge* et al. 1983).

Airway defence reflexes evoked by administration of prostaglandins have been studied in experimental animals and man. Prostaglandin $F_{2\alpha}$, injected intravenously (3 μg kg^{-1}), causes rapid, shallow breathing and bronchoconstriction in dogs (*Wasserman* 1975). A prominent component of the bronchoconstriction is of vagal reflex origin and is abolished by atropine. The bronchodilator prostaglandins have irritant effects on the airways of man (*Herxheimer* and *Roetscher* 1971; *Kawakami* et al. 1973) and occasionally when delivered as an aerosol (*Kawakami* et al. 1973; *Smith* and *Cuthbert* 1976) or as an intravenous infusion (*Smith* 1973) have paradoxical bronchoconstrictor effects that may also be of vagal reflex origin. The bronchoconstrictor prostaglandin $F_{2\alpha}$, in the doses employed in reflex experiments (*Wasserman* 1975), stimulates irritant receptors in dogs as well as bronchial and pulmonary C fibres, but the bronchodilator prostaglandins have only minor effects on irritant receptors and appear to evoke irritant and bronchoconstrictor reflex effects from the lower airways mainly by stimulating afferent C fibres (*Coleridge* et al. 1976). Reflex tracheal contraction is evoked in dogs when bronchodilator prostaglandins are injected into the right atrium or, in much smaller doses, directly into a bronchial artery; it is abolished by vagotomy or by administration of atropine (*Roberts* et al. 1981a).

6.2.3 Relative Roles of C Fibres and Irritant Receptors

We speculate that the afferent C fibre innervation of the lower respiratory tract not only plays a greater role in the airway defence reflexes than was formerly acknowledged but also accounts for the rapid shallow breathing characteristic of airway defence reflexes in experimental animals. Because irritant (rapidly adapting) receptors appear to increase inspiration and to elevate inspiratory off-switch threshold (*Knowlton* and *Larrabee* 1946; *Glogowska* et al. 1972), and because repeated stimulation of these receptors by increased airflow (*Pack* 1981) fails to shorten T_E (*Bartoli* et al. 1975; *Pack* et al. 1981), it seems doubtful at present that input from this particular group of myelinated afferents, once believed to account for the ventilatory effects of histamine and other airway irritants, can indeed function as a trigger for rapid shallow breathing, a pattern of breathing that is more typical of what is known of the effects of stimulating afferent C fibres. Irritant receptors undoubtedly play a part in the airway defence reflexes, causing the characteristic increased inspiratory efforts and gasps or sighs and contributing to the cough reflex, but their participation in the reflex bronchoconstriction and increased airway secretion, though likely, has not been established experimentally.

Of the afferent C fibres so far identified in the lower respiratory tract, those supplying the conducting airways seem likely to play the major part in the airway defence reflexes in most circumstances; they are highly sensitive to chemicals, and are strategically located to respond to chemicals that are inhaled and to autocoids that are released in the bronchial walls. Because they resemble in so many ways the cutaneous C fibres that mediate sensations of itch and burning pain in inflammation of the skin, it is not unreasonable to suppose that they mediate the rather similar sensations arising from the lower respiratory tract. Neurally mediated inflammatory phenomena are readily induced by application of irritants to the skin and are accompanied by hyperalgesia (*Lynn* 1977). A somewhat similar phenomenon in the airway mucosa may account for the hyper-reflexia described after exposure to ozone (*Lee* et al. 1979) and after respiratory tract infections (*Dixon* et al. 1979b) in dogs. In the skin chemosensitive C fibres are believed to be responsible for the spreading flare of the so-called axon reflex (reviewed by *Chapman* and *Goodell* 1964). *Lundberg* and *Saria* (1982) recently described what they suggested was a similar phenomenon of antidromic vasodilatation in the tracheal mucosa of rats, evoked by stimulation of the peripheral cut end of the vagus nerve after efferents had been blocked by atropine and hexamethonium. If the presence of such 'axon reflexes' in the lower airways is confirmed, then the chemosensitive airway C fibres are the prime candidates for a causative role.

Although pulmonary C fibres are stimulated by a variety of foreign irritants, including the common air pollutant SO_2, they appear to be insensitive to certain lung autocoids, and in any event are in a region of lung tissue richly supplied with the enzymes that break down many lung autocoids (*Junod* 1975, 1977). Moreover, when irritants are inhaled some proportion is likely to be trapped at bronchial bifurcations and diluted by secretion before it can reach the more distal lung divisions. Hence pulmonary C fibres seem likely to play a role subsidiary to that of airway C fibres in triggering airway defence reflexes. However, prolonged exposure to irritants undoubtedly recruits activity in pulmonary as well as bronchial C fibres, and the profound cardiovascular depressor effects often seen when experimental animals are exposed to powerful airway irritants are probably a sign that pulmonary C fibres have been recruited.

6.3 Role of C Fibres in Lung Disease

That afferent C fibres function as lower respiratory tract nociceptors is generally accepted, and there is little that can be added to the preceding sections regarding their reflex role in lung disease. Models of lung disease in experimental animals provided early indications that afferent C fibres were stimulated: thus the familiar pattern of apnoea, rapid shallow breathing, bradycardia and hypotension, abolished or greatly reduced by vagotomy, was induced by acute, severe pulmonary congestion in cats (*Churchill* and *Cope* 1929) and dogs (*Aviado* et al. 1951; *Downing* 1957), by pulmonary embolism in several species (*Whitteridge* 1950; *Cahill* et al. 1961) and by pulmonary anaphylaxis in rabbits (*Karczewski* and *Widdicombe* 1969b).

Apart from such changes in breathing and heart rate, the contribution of vagal reflexes to the abnormalities of lung disease may be difficult to assess, for changes in blood gases are often a complicating factor, and the pathological process often causes release of autocoids that have powerful direct actions on bronchial and vascular smooth muscle. Disease may also bring about structural changes in the lungs and airways. Nevertheless, a characteristic respiratory abnormality of experimental models of lung disease in animals is a vagally mediated tachypnoea with shortening of T_I and T_E, and reduction in V_T, a breathing pattern that is equally characteristic of the airway defence reflex induced by stimulation of lower respiratory C fibres by chemicals (see above). Tachypnoea is also common in disease of the lungs and airways in man (*Guz* 1977a).

Administration of histamine (see above) or antigen (*Karczewski* and *Widdicombe* 1969c; *Gold* et al. 1972; *Kessler* et al. 1973; *Cotton* et al. 1977) to the lower airways of animals is often regarded as providing a useful experimental model of human asthma (*Widdicombe* 1977b), and although not all authorities would subscribe to this view, it has enabled the identification of possible vagal reflex components of asthma and of allied bronchial hyper-reactivity syndromes. Such topics have been the subject of recent reviews (*Nadel* 1976; *Widdicombe* 1979; *Boushey* et al. 1980). A role for bronchial C fibres in the tachypnoea and other vagal reflex components of human asthma can be postulated on the basis of their sensitivity to autocoids such as histamine, bradykinin and the prostaglandins (see above), which are known to be released in asthma (*Nakano* and *Rogers* 1976; *Garcia Leme* 1978; *Plaut* and *Lichtenstein* 1978), and from what is known of their reflex properties. Bradykinin has been found in concentrations of $1-12$ ng ml^{-1} in forearm mixed venous blood in patients with severe asthma (*Abe* et al. 1967); hence concentrations at the site of release in the lung may be much higher. Injection of bradykinin in doses of only $0.002-1.5$ µg into the bronchial artery of dogs is effective in evoking a reflex increase in airway smooth muscle tone and submucosal gland secretion (*Roberts* et al. 1981b; *Davis* et al. 1982b), and slow infusion of equally small amounts evokes rapid shallow breathing (*Coleridge* et al. 1983). It seems likely that the small amounts of bradykinin administered to evoke these reflex effects produced concentrations in the vicinity of the vagal endings in the bronchial wall within the range of those reached during endogenous bradykinin production under pathological conditions.

Neurophysiological studies in dogs, cats and rabbits confirm that afferent C fibres supplying the lower respiratory tract are stimulated by pulmonary congestion (*Coleridge* and *Coleridge* 1975, 1977a), embolism (*Paintal* et al. 1973; *Armstrong* et al. 1976; *Armstrong* and *Miller* 1980) and inflammation (*Frankstein* and *Sergeeva* 1966). The results of reflex experiments suggest that the tachypnoea of these pathological conditions is due mainly to increased activity in this C fibre pathway. The tachypnoea of pulmonary embolism, for example, is described as unaffected by oxygen breathing and as persisting at low vagal temperatures although abolished by vagotomy (*Whitteridge* 1950). In rabbits the tachypnoea of pulmonary embolism was found to be unaffected by anodal polarization block of the myelinated fibres in the vagus and was thought to be due entirely to activity in afferent vagal C fibres (*Guz* and *Trenchard* 1971). The tachypnoea induced by lung inflammation in rabbits could not be ascribed to an increase in body temperature or to changes in blood gases; again it was dependent on vagal conduction, and although attenuated it survived block of conduction in myelinated vagal fibres (*Trenchard* et al. 1972).

Inflammation of lung lobes has been induced in animals with a view to obtaining some insight into the mechanism of the dyspnoea, tachypnoea and exercise limitation observed in diffuse pulmonary disease in man. There seems little doubt that in man these respiratory abnormalities are to a large extent of vagal afferent origin. Thus *Guz* et al. (1970) have shown that breath-holding time is often greatly reduced in patients with lung disease, even when arterial PCO_2 is below normal and arterial PO_2 is elevated by oxygen breathing. Bilateral vagal blockade with local anaesthetic greatly increased breath-holding time and diminished tachypnoea in such patients, and in a patient whose disease was confined to one lung, section of the vagus on the side of the lesion not only increased breath-holding time but also produced a maintained improvement of exertional dyspnoea. Many patients with lung disease have little or no respiratory discomfort at rest, but have pronounced dyspnoea and tachypnoea during even mild exercise (*Guz* et al. 1970; *Guz* 1977b).

Frankstein and *Sergeeva* (1966) and *Frankstein* (1970) suggested, on the basis of experiments in cats, that tonically increased input in afferent C fibres from the damaged lung increases the excitability of the central mechanisms responsible for respiratory timing and reduces the influence of inhibitory signals. Thus in cats with inflammation of one lung the Hering-Breuer inhibition of breathing evoked by lung inflation was greatly reduced, a reduction that was equally pronounced whether the normal or the inflamed lung was inflated (*Frankstein* 1970). Such a mechanism probably accounts for the results of *Phillipson* et al. (1975b) in dogs. These investigators found an abnormal increase in breathing rate and limitation of V_T, together with decreased end-tidal PCO_2 and increased arterial pH, in dogs exercising during the recovery phase of experimental pneumonitis, at a time when resting values for V_T and breathing rate had returned to normal levels (Fig. 31). Complete vagal blockade diminished the excitatory response to exercise, and increased exercise tolerance in some cases.

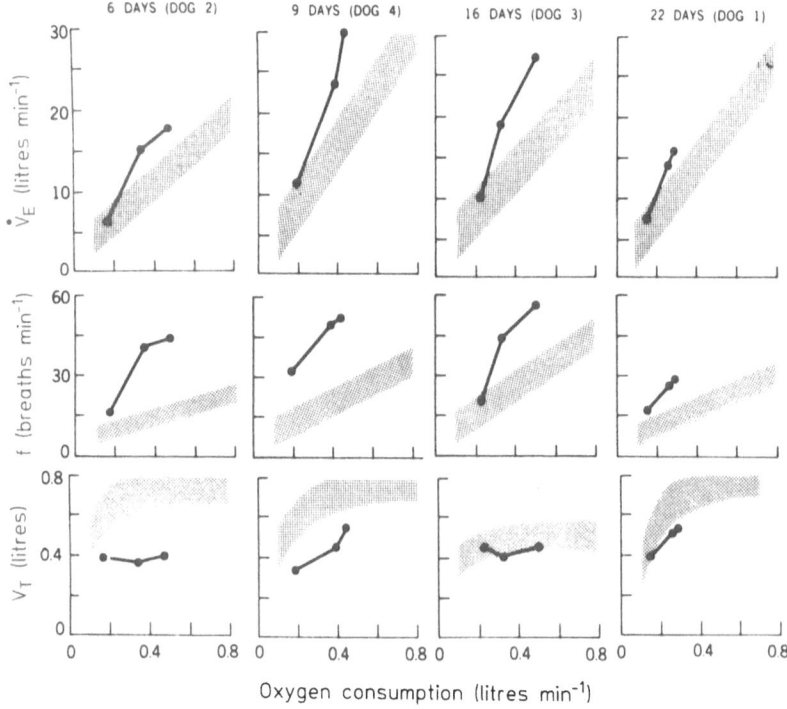

Fig. 31. Effect of experimental pneumonitis on the ventilatory response of four dogs to increasing levels of exercise. V_E, ventilation volume (litres min^{-1}); f, respiratory frequency (breaths min^{-1}); V_T, tidal volume (litres); increasing levels of exercise expressed as O_2 consumption (litres min^{-1}). Diffuse interstitial pneumonitis was induced by intravenous administration of complete Freund's adjuvant; above, days after administration of adjuvant. *Shaded areas* represent 95% confidence limits of all control data obtained in each dog when healthy. First data point (i.e. lowest O_2 consumption) for each dog represents values obtained standing at rest. (*Phillipson* et al. 1975b)

To interpret the respiratory abnormalities of lung disease solely in terms of an increase in input in afferent vagal C fibres would clearly be an oversimplification, for activity in myelinated afferent vagal fibres, and probably also in the sympathetic afferent pathway, is likely to be greatly altered in these circumstances. If we consider only the vagal afferent pathway, since the tachypnoea of lung disease is known to be a vagal effect, the discharge of slowly adapting pulmonary stretch receptors has been found to increase throughout the respiratory cycle in pulmonary congestion (*Marshall* and *Widdicombe* 1958; *Costantin* 1959) and to a smaller extent in embolism (*Armstrong* et al. 1976, 1979). In inspiration the increased discharge of slowly adapting pulmonary stretch receptors will combine with the tachypnoeic influence of afferent C fibres in limiting V_T and T_I (*Bradley* 1977; *Trenchard* 1977); in expiration any increased discharge of slowly adapting stretch receptors will tend to lengthen, rather shorten, T_E (*Bartoli* et al. 1973; *Trenchard* 1977); hence at this phase the shortening of T_E must be ascribed to the dominant influence of afferent C fibres.

Irritant or rapidly adapting receptors are strongly stimulated by congestion (*Sellick* and *Widdicombe* 1969; *Mills* et al. 1970) and, to a lesser extent, by embolism (*Armstrong* et al. 1976, 1979), and often discharge bursts of impulses with each inflation; from what is known of their reflex properties they are likely to increase the vigour of inspiratory efforts in these conditions (*Glogowska* et al. 1972). The changes in slowly and rapidly adapting receptor activity are thought to result from increased lung stiffness; the factors responsible for the increased stiffness are acknowledged to be exceedingly complex and to include release of chemicals which may further sensitize the nerve endings. Similar mechanical and chemical factors are likely to operate in inflammatory disease of the lung, and it seems reasonable to agree with *Phillipson* et al. (1975b) that an overall increase in the activity of slowly and rapidly adapting receptors will contribute to the abnormalities of breathing pattern in this condition also. There is some evidence, however, that the phasic activity of vagal mechanoreceptors with myelinated fibres actually decreases in severely inflamed regions of the lung (*Frankstein* and *Sergeeva* 1966).

7 Conclusions

We have attempted in this review to give an account of an afferent vagal input from the lower respiratory tract that has still to be explored fully, and to present experimental evidence that this fine fibre afferent system plays a significant role in the neural control of respiratory function in both normal and pathological circumstances.

We have made a distinction between the afferent C fibres that innervate the lung parenchyma adjacent to the pulmonary capillary bed and those that innervate the conducting airways, even though the afferent C fibres in the two locations appear to have reflex properties that are at least qualitatively similar. We believe that the functional significance of these lower respiratory tract C fibres is determined not simply by their location but also by certain differences in afferent properties that should not be overlooked.

Douglas and *Ritchie* (1962) suggested, in their review of mammalian non-myelinated nerve fibres, that the teleological advantage of the fine-fibre afferent system, especially in the case of a visceral input where speed of impulse transmission was not of primary importance, was that it allows fibres of a variety of sensory modalities to be accommodated in a small cross-sectional area of nerve trunk. There is no reason to think that the full range of sensory modalities of the afferent C fibres in the lungs and airways has yet been explored. The custom of injecting certain chemicals to identify the presence of lower respiratory tract C fibres when recording the activity in vagal strands is highly selective, so that even now our view of this afferent fibre system may be unnecessarily narrow.

Some of the conclusions arrived at in these pages are either purely speculative or derived from experimental evidence that is at best indirect. Whether they prove to be correct or incorrect — and some are sure to fall into the latter category — their purpose will have been served if they are put to the test of experiment.

Acknowledgements. This work was supported by Program Project Grant HL-24136 from the National Heart, Lung, and Blood Institute. We are indebted to Ms. Rolinda Wang for typing the manuscript and to Mr. Albert Dangel for photographing the illustrations.

References

Abbott BC, Howarth JV, Ritchie JM (1965) The initial heat production associated with the nerve impulse in crustacean and mammalian non-myelinated nerve fibres. J Physiol (Lond) 178:368–383

Abe K, Watanabe N, Kumagi N, Mouri T, Seki T, Yoshinaga K (1967) Circulating plasma kinin in patients with bronchial asthma. Experientia 23:626–627

Adrian ED (1933) Afferent impulses in the vagus and their effect on respiration. J Physiol (Lond) 79:332–358

Agostoni E, Chinnock JE, Daly M de B, Murray JG (1957) Functional and histological studies of the vagus nerve and its branches to the heart, lungs and abdominal viscera in the cat. J Physiol (Lond) 135:182–205

Ahluwalia J, Devanandan MS, Shukla SB (1977) Some functional properties of the inhibition of skeletal muscles by activation of type-J pulmonary endings. In: Paintal AS, Gill-Kumar P (eds) Krogh centenary symposium on respiratory adaptations, capillary exchange and reflex mechanisms. Vallabhbhai Patel Chest Inst, Delhi, pp 447–465

Allison DJ, Clay TP, Hughes JMB, Jones HA, Shevis A (1974) Effects of nasal stimulation on total respiratory resistance in the rabbit. J Physiol (Lond) 239:23–24P

Anand A, Paintal AS (1980) Reflex effects following selective stimulation of J receptors in the cat. J Physiol (Lond) 299:553–572

Anand A, Loeschcke HH, Marek W, Paintal AS (1982) Significance of the respiratory drive by impulses from J receptors. J Physiol (Lond) 325:14P

Anrep GV, Pascual W, Rössler R (1936) Respiratory variations of the heart rate I. The reflex mechanism of the respiratory arrhythmia. Proc R Soc Lond [Biol] 119: 191–217

Armstrong DJ, Luck JC (1974) A comparative study of irritant and type J receptors in the cat. Respir Physiol 21:47–60

Armstrong DJ, Miller SA (1980) Lung irritant and C fibre responses to embolism in thrombocytopaenic rabbits. J Physiol (Lond) 303:41–42P

Armstrong DJ, Miller SA (1981) Intrapulmonary C fibres contribute to the afferent arm of the vagally mediated responses to right atrial injections of prostacyclin into rabbits. J Physiol (Lond) 310:65–66P

Armstrong DJ, Luck JC, Martin VM (1976) The effect of emboli upon intrapulmonary receptors in the cat. Respir Physiol 26:41–54

Armstrong DJ, Lacey M, Luck JC (1979) The role of platelets in the post-embolic responses of slowly adapting intrapulmonary stretch receptors in the rabbit. J Physiol (Lond) 289:93–94P

Aviado DM, Li TH, Kalow W, Schmidt CF, Turnbull GL, Peskin GW, Hess ME, Weiss AJ (1951) Respiratory and circulatory reflexes from the perfused heart and pulmonary circulation of the dog. Am J Physiol 165:261–277

Baker DG, Coleridge HM, Coleridge JCG (1979) Vagal afferent C fibres from the ventricle. In: Hainsworth R, Kidd C, Linden RJ (eds) Cardiac receptors. Cambridge University Press, Cambridge, pp 117–137

Banister J, Fegler G, Hebb C (1950) Initial respiratory responses to the intratracheal inhalation of phosgene or ammonia. Q J Exp Physiol 35:233–250

Barer GR, Nusser E (1953) The part played by bronchial muscles in pulmonary reflexes. Br J Pharmacol 8:315–320

Barer GR, Nusser E (1958) Cardiac output during excitation of chemoreflexes in the cat. Br J Pharmacol 13:372–377

Bartoli A, Bystrzycka E, Guz A, Jain SK, Noble MIM, Trenchard D (1973) Studies of the pulmonary vagal control of central respiratory rhythm in the absence of breathing movements. J Physiol (Lond) 230:449–465

Bartoli A, Cross BA, Guz A, Huszczuk A, Jefferies R (1975) The effect of varying tidal volume on the associated phrenic motoneurone output: studies of vagal and chemical feedback. Respir Physiol 25:135–155

Bevan JA (1962) Action of lobeline and capsaicine on afferent endings in the pulmonary artery of the cat. Circ Res 10:792–797

Bevan JA, Murray JF (1963) Evidence for a ventilation modifying reflex from the pulmonary circulation in man. Proc Soc Exp Biol Med 114:393–396

Bleecker ER, Cotton DJ, Fischer SP, Graf PD, Gold WM, Nadel JA (1976) The mechanism of rapid, shallow breathing after inhaling histamine aerosol in exercising dogs. Am Rev Respir Dis 114:909–916

Boushey HA, Richardson PS, Widdicombe JG (1972) Reflex effects of laryngeal irritation on the pattern of breathing and total lung resistance. J Physiol (Lond) 224:501–513

Boushey HA, Holtzman MJ, Sheller JR, Nadel JA (1980) Bronchial hyperreactivity. Am Rev Respir Dis 121:389–413

Bouverot P, Crance JP, Dejours P (1970) Factors influencing the intensity of the Breuer-Hering inspiration-inhibiting reflex. Respir Physiol 8:376–384

Bradley GW (1977) Control of the breathing pattern. In: Widdicombe JG (ed) International review of physiology. University Park Press, Baltimore, pp 185–217, (Respiratory Physiology II, vol 14)

Bradley GW, Euler C von, Marttila I, Roos B (1974) Transient and steady state effects of CO_2 on mechanisms determining rate and depth of breathing. Acta Physiol Scand 92:341–350

Brender D, Webb-Peploe MM (1969) Vascular responses to stimulation of pulmonary and carotid baroreceptors by capsaicin. Am J Phyisol 217:1837–1845

Brodie TG (1900) The immediate action of an intravenous injection of blood-serum. J Physiol (Lond) 26:48–71

Brodie TG, Russell AE (1900) On reflex cardiac inhibition. J Physiol (Lond) 26:92–106

Brown JK, Leff AR, Frey MJ, Reed BR, Gold WM (1980) Physiological and pharmacological properties of canine trachealis muscle in vivo. J Appl Physiol 49:84–94

Bucher K (1958) Pathophysiology and pharmacology of cough. Pharmacol Rev 10: 43–58

Cahill JM, Attinger EO, Byrne JJ (1961) Ventilatory responses to embolization of lung. J Appl Physiol 16:469–472

Casey KL, Blick M (1969) Observations on anodal polarisation of cutaneous nerve. Brain Res 13:155–167

Cassidy SS, Mitchell JH (1981) Effects of positive pressure breathing on right and left ventricular preload and afterload. Fed Proc 40:2178–2181

Cassidy SS, Robertson CH, Pierce AK, Johnson RL (1978) Cardiovascular effects of positive end-expiratory pressure in dogs. J Appl Physiol 44:743–750

Cassidy SS, Eschenbacher WL, Johnson RL (1979) Reflex cardiovascular depression during unilateral lung hyperinflation in the dog. J Clin Invest 64:620–626

Chapman LF, Goodell H (1964) The participation of the nervous system in the inflammatory reaction. Ann NY Acad Sci 116:990–1017

Churchill ED, Cope O (1929) The rapid shallow breathing resulting from pulmonary congestion and edema. J Exp Med 49:531–537

Clark FJ, Euler C von (1972) On the regulation of depth and rate of breathing. J Physiol (Lond) 222:267–295

Coleridge HM, Coleridge JCG (1975) Two types of afferent vagal C-fibre in the dog lung: their stimulation by pulmonary congestion. Fed Proc 34:372

Coleridge HM, Coleridge JCG (1977a) Afferent vagal C-fibers in the dog lung: their discharge during spontaneous breathing and their stimulation by alloxan and pulmonary congestion. In: Paintal AS, Gill-Kumar P (eds) Krogh centenary symposium on respiratory adaptations, capillary exchange and reflex mechanisms. Vallabhbhai Patel Chest Inst, Delhi, pp 393–406

Coleridge HM, Coleridge JCG (1977b) Impulse activity in afferent vagal C-fibres with endings in the intrapulmonary airways of dogs. Respir Physiol 29:125–142

Coleridge HM, Coleridge JCG (1981) Afferent fibers involved in defense reflexes from the respiratory tract. In: Hutas I, Debreczeni LA (eds) Advances in physiological science, vol 10. Academiai Kiado, Budapest, pp 467–477

Coleridge HM, Coleridge JCG (to be published) Reflexes from the tracheobronchial tree and lungs. In: Cherniak N, Widdicombe JG (eds) Control of breathing. American Physiological Society, Washington DC (Handbook of physiology, The respiratory system, vol 1)

Coleridge HM, Coleridge JCG, Kidd C (1964a) Cardiac receptors in the dog, with particular reference to two types of afferent ending in the ventricular wall. J Physiol (Lond) 174:323–339

Coleridge HM, Coleridge JCG, Kidd C (1964b) Role of the pulmonary arterial baroreceptors in the effects produced by capsaicin in the dog. J Physiol (Lond) 170: 272–285

Coleridge HM, Coleridge JCG, Luck JC (1965) Pulmonary afferent fibres of small diameter stimulated by capsaicin and by hyperinflation of the lungs. J Physiol (Lond) 179:248–262

Coleridge HM, Coleridge JCG, Luck JC, Norman J (1968) The effect of four volatile anesthetic agents on the impulse activity of two types of pulmonary receptor. Br J Anaesth 40:484–492

Coleridge HM, Coleridge JCG, Dangel A, Kidd C, Luck JC, Sleight P (1973a) Impulses in slowly conducting vagal fibers from afferent endings in the veins, atria and arteries of dogs and cats. Circ Res 33:87–97

Coleridge HM, Coleridge JCG, Rosenthal F, Dangel A (1973b) Stimulation of C-fibers accompanying anodal polarization block of A-fibers in the vagus nerves of cats. Fed Proc 32:355

Coleridge HM, Coleridge JCG, Ginzel KH, Baker DG, Banzett RB, Morrison MA (1976) Stimulation of 'irritant' receptors and afferent C-fibres in the lungs by prostaglandins. Nature 264:451–453

Coleridge HM, Coleridge JCG, Baker DG, Ginzel KH, Morrison MA (1978a) Comparison of the effects of histamine and prostaglandin on afferent C-fiber endings and
irritant receptors in the intrapulmonary airways. In: Fitzgerald RS, Gautier H,
Lahiri S (eds) Regulation of respiration during sleep and anesthesia. Adv Exp Med
Biol 99:291–305

Coleridge HM, Coleridge JCG, Banzett RB (1978b) Effect of CO_2 on afferent vagal
endings in the canine lung. Respir Physiol 34:135–151

Coleridge HM, Roberts AM, Coleridge JCG (1981) Stimulation of bronchial C-fibers
in dogs causes rapid shallow breathing. Fed Proc 40:452

Coleridge HM, Roberts AM, Coleridge JCG (1982b) Effect of vagal cooling on activity in lung afferent fibers in dogs. Fed Proc 41:986

Coleridge HM, Coleridge JCG, Roberts AM (1983) Rapid shallow breathing evoked by
selective stimulation of airway C fibres in dogs. J Physiol (Lond) 340:415–433

Coleridge JCG, Coleridge HM (1977c) Afferent C-fibers and cardiorespiratory chemoreflexes. Am Rev Respir Dis 115:251–260

Coleridge JCG, Coleridge HM (1979) Chemoreflex regulation of the heart. In: Berne
RM (ed) The cardiovascular system, vol 1. The heart. American Physiological
Society, Bethesda, pp 653–676 (Handbook of physiology, vol 2)

Coleridge JCG, Coleridge HM, Roberts AM, Kaufman MP, Baker DG (1982a) Tracheal
contraction and relaxation initiated by lung and somatic afferents in dogs. J Appl
Physiol 52:984–990

Costantin LL (1959) Effect of pulmonary congestion on vagal afferent activity. Am J
Physiol 196:49–53

Cotton DJ, Bleecker ER, Fischer SP, Graf PD, Gold WM, Nadel JA (1977) Rapid,
shallow breathing after ascaris suum antigen inhalation: role of vagus nerves. J Appl
Physiol 42:101–106

Cromer SP, Young RH, Ivy AC (1933) On the existence of afferent respiratory impulses mediated by the stellate ganglia. Am J Physiol 104:468–475

Crosfill ML, Widdicombe JG (1961) Physical characteristics of the chest and lungs and
the work of breathing in different mammalian species. J Physiol (Lond) 158:1–14

Cross BA, Guz A, Jain SK, Archer S, Stevens J, Reynolds F (1976) The effect of
anaesthesia of the airway in dog and man: a study of respiratory reflexes, sensations
and lung mechanics. Clin Sci 50:439–454

Dain DS, Boushey HA, Gold WM (1975) Inhibition of respiratory reflexes by local
anesthetic aerosols in dogs and rabbits. J Appl Physiol 38:1045–1050

Daly M de B (1972) Interaction of cardiovascular reflexes. Lectures Scient Basis Med,
The Scientific Basis of Medicine Annual Reviews, University of London, Athlone
Press, London, pp 307–332

Daly I de B, Hebb C (1966) Pulmonary and bronchial vascular systems. Arnold,
London

Daly M de B, Robinson BH (1968) An analysis of the reflex systemic vasodilator response elicited by lung inflation in the dog. J Physiol (Lond) 195:387–406

Daly M de B, Hazzledine JL, Ungar A (1967) The reflex effects of alterations in lung
volume on systemic vascular resistance in the dog. J Physiol (Lond) 188:331–351

Davies A, Dixon M, Callanan D, Huszczuk A, Widdicombe JG, Wise JCM (1978) Lung
reflexes in rabbits during pulmonary stretch receptor block by sulphur dioxide.
Respir Physiol 34:83–101

Davis B, Chinn R, Gold J, Popovac D, Widdicombe JG, Nadel JA (1982a) Hypoxemia
reflexly increases secretion from tracheal submucosal glands in dogs. J Appl Physiol
53:1416–1419

Davis B, Roberts AM, Coleridge HM, Coleridge JCG (1982b) Reflex tracheal gland
secretion evoked by stimulation of bronchial C-fibers in dogs. J Appl Physiol 53:
985–991

Dawes GS, Comroe JH Jr (1954) Chemoreflexes from the heart and lungs. Physiol Rev
34:167–201

Dawes GS, Mott JC, Widdicombe JG (1951) Respiratory and cardiovascular reflexes from the heart and lungs. J Physiol (Lond) 115:258–291

Dawes GS, Mott JC, Widdicombe JG (1952) Chemoreceptor reflexes in the dog and the action of phenyl diguanide. Arch Int Pharmacodyn Ther 90:203–222

DeKock MA, Nadel JA, Zwi S, Colebatch HJH, Olsen CR (1966) New method for perfusing bronchial arteries: histamine bronchoconstriction and apnea. J Appl Physiol 21:185–194

Delpierre S, Grimaud C, Jammes Y, Mei N (1981) Changes in activity of vagal broncho-pulmonary C fibres by chemical and physical stimuli in the cat. J Physiol (Lond) 316:61–74

Deshpande SS, Devanandan MS (1970) Reflex inhibition of monosynaptic reflexes by stimulation of type J pulmonary endings. J Physiol (Lond) 206:345–357

Dickinson CJ, Paintal AS (1970) Stimulation of type-J pulmonary receptors in the cat by carbon dioxide. Clin Sci 38:33P

Dixon M, Jackson DM, Richards IM (1979a) The effects of histamine, acetylcholine and 5-hydroxytryptamine on lung mechanics and irritant receptors in the dog. J Physiol (Lond) 287:393–403

Dixon M, Jackson DM, Richards IM (1979b) The effect of a respiratory tract infection on histamine-induced changes in lung mechanics and irritant receptor discharge in dogs. Am Rev Respir Dis 120:843–848

Douglas WW, Ritchie JM (1957) On excitation of non-medullated afferent fibres in the vagus and aortic nerves by pharmacological agents. J Physiol (Lond) 138:31–43

Douglas WW, Ritchie JM (1962) Mammalian nonmyelinated nerve fibers. Physiol Rev 42:297–334

Douglas WW, Ritchie JM, Schaumann W (1956) Depressor reflexes from medullated and non-medullated fibres in the rabbit's aortic nerve. J Physiol (Lond) 132:187–198

Downing SE (1957) Reflex effects of acute hypertension in the pulmonary vascular bed of the dog. Yale J Biol Med 30:43–56

During M von, Andres KH, Irvani J (1974) The fine structure of the pulmonary stretch receptor in the rat. Z Anat Entwickl-Gesch 143:215–222

Eckenhoff JE, Comroe JH Jr (1951) Blocking action of tetra-ethylammonium on lobelin-induced thoracic pain. Proc Soc Exp Biol Med 76:725–726

Elftman AG (1943) The afferent and parasympathetic innervation of the lungs and trachea of the dog. Am J Anat 72:1–28

Evans MH, McPherson A (1958) The effects of stimulation of visceral afferent nerve fibres on somatic reflexes. J Physiol (Lond) 140:201–212

Fastier FN, McDowall MA, Wall H (1959) Pharmacological properties of phenyl-diguanide and other amidine derivatives in relation to those of 5-hydroxytrypt-amine. Br J Pharmacol 14:527–535

Fillenz M, Widdicombe JG (1972) Receptors of the lungs and airways. In: Neil E (ed) Enteroceptors. Springer, Berlin Heidelberg New York, pp 81–112 (Handbook of sensory physiology, vol 3/1)

Fishman NH, Phillipson EA, Nadel JA (1973) Effect of differential vagal cold blockade on breathing pattern in conscious dogs. J Appl Physiol 34:754–758

Fox B, Bull TB, Guz A (1980) Innervation of alveolar walls in the human lung: an electron microscopic study. J Anat (Lond) 131:683–692

Frankstein SI (1970) Neural control of respiration. In: Porter R (ed) Breathing: Hering-Breuer centenary symposium. Churchill, London, pp 53–58

Frankstein SI, Sergeeva ZN (1966) Tonic activity of lung receptors in normal and pathological states. Nature 210:1054–1055

Franz DN, Iggo A (1968) Conduction failure in myelinated and non-myelinated axons at low temperatures. J Physiol (Lond) 199:319–345

Franz DN, Perry RS (1974) Mechanisms for differential block among single myelinated and non-myelinated axons by procaine. J Physiol (Lond) 236:193–210

Gallagher JT, Kent PW, Passatore M, Phipps RJ, Richardson PS (1975) The composition of tracheal mucus and the nervous control of its secretion in the cat. Proc R Soc Lond [Biol] 192:49−76

Garcia Leme J (1978) Bradykinin-System. In: Vane JR, Ferreira SH (eds) Inflammation. Springer, Berlin Heidelberg New York, pp 464−522 (Handbook of experimental pharmacology, vol 50/1)

Ginzel KH (1978) The respiratory effect of vagal lung afferents ('J receptors') excited by phenyldiguanide (PDG). Fed Proc 37:579

Ginzel KH, Eldred E (1969) Relief of decerebrate rigidity by viscero-somatic reflex action. Proc West Pharmacol Soc 12:41−44

Ginzel KH, Eldred E (1977) Reflex depression of somatic motor activity from heart, lungs and carotid sinus. In: Paintal AS, Gill-Kumar P (eds) Krogh centenary symposium on respiratory adaptations, capillary exchange and reflex mechanisms. Vallabhbhai Patel Chest Institute, Delhi, pp 358−394

Ginzel KH, Eldred E, Sasaki Y (1969) Comparative study of the actions of nicotine and succinylcholine on the monosynaptic reflex and spindle afferent activity. Int J Neuropharmacol 8:515−533

Ginzel KH, Eldred E, Watanabe S, Grover F (1971) Drug induced depression of gamma afferent activity. III. Viscerosomatic reflex action of phenyldiguanide, veratridine and 5-hydroxytryptamine. Neuropharmacology 10:77−91

Ginzel KH, Morrison MA, Baker DG, Coleridge HM, Coleridge JCG (1978) Stimulation of afferent vagal endings in the intrapulmonary airways by prostaglandin endoperoxide analogues. Prostaglandins 15:131−138

Glick G, Wechsler AS, Epstein SE (1969) Reflex cardiovascular depression produced by stimulation of pulmonary stretch receptors in the dog. J Clin Invest 48:467−473

Glogowska M, Richardson PS, Widdicombe JG, Winning AJ (1972) The role of the vagus nerves, peripheral chemoreceptors and other afferent pathways in the genesis of augmented breaths in cats and rabbits. Respir Physiol 16:179−196

Gold WM, Kessler GF, Yu DYC (1972) Role of vagus nerves in experimental asthma in allergic dogs. J Appl Physiol 33:719−725

Green JF, Sheldon MI (1983) Ventilatory changes associated with changes in pulmonary blood flow in dogs. J Appl Physiol 54:997−1002

Greenwood PV, Hainsworth R, Karim F, Morrison GW, Sofola OA (1977) Cardiac inotropic responses from lung inflation. J Physiol (Lond) 271:37−38P

Guz A (1977a) Control of ventilation in man with special reference to abnormalities in asthma. In: Lichtenstein LM, Austen KF (eds) Asthma. Physiology, immunopharmacology, and treatment. Academic Press, New York, pp 211−223

Guz A (1977b) Respiratory sensations in man. Br Med Bull 33:175−177

Guz A, Trenchard DW (1971) The role of non-myelinated vagal afferent fibres from the lungs in the genesis of tachypnoea in the rabbit. J Physiol (Lond) 213:345−371

Guz A, Noble MIM, Widdicombe JG, Trenchard D, Muschin WW (1966) The effect of bilateral block of vagus and glossopharyngeal nerves on the ventilatory response to CO_2 of conscious man. Respir Physiol 1:206−210

Guz A, Noble MIM, Eisele JH, Trenchard D (1970) Experimental results of vagal block in cardiopulmonary disease. In: Porter R (ed) Breathing: Hering-Breuer centenary symposium. Churchill, London, pp 315−329

Hahn HL, Johnson HG, Chow AW, Graf PD, Nadel JA (1982a) Reflex stimulation of submucosal gland secretion by sulfur dioxide. Fed Proc 41:1509

Hahn HL, Sasaki K, Nadel JA (1982b) Effects of sulfur dioxide on ventilation in conscious dogs. Am Rev Respir Dis 125:220

Hainsworth R (1974) Circulatory responses from lung inflation in anesthetized dogs. Am J Physiol 226:247−255

Halmagyi DFJ, Colebatch HJH (1961) Cardiorespiratory effects of experimental lung embolism. J Clin Invest 40:1785−1796

Hammouda M, Wilson WH (1935a) Further observations on the respiratory-accelerator fibres of the vagus. J Physiol (Lond) 85:62–72

Hammouda M, Wilson WH (1935b) The presence in the vagus of fibres transmitting impulses augmenting the frequency of respiration. J Physiol (Lond) 83:292–312

Hammouda M, Wilson WH (1939) Reflex acceleration of respiration arising from excitation of the vagus or its terminations in the lungs. J Physiol (Lond) 94:497–524

Hanacek J, Widdicombe JG, Korpas J (1980) Stretch receptors of the lung and their effect on lung defensive reflexes. Proceedings of the 28th international congress of physiology, Budapest, vol XIV. Hungarian Physiological Society, Budapest, pp 460

Head H (1889) On the regulation of respiration, part I. Experimental. J Physiol (Lond) 10:1–70

Hendrix SG, Munoz NM, Leff AR (1983) Physiological and pharmacological response of canine bronchial smooth muscle in situ. J Appl Physiol 54:215–224

Hering E, Breuer J (1868) Die Selbststeuerung der Athmung durch den Nervus vagus. Sitzber Akad Wiss, Wien 57:672–677

Herxheimer H, Roetscher I (1971) Effects of prostaglandin E_1 on lung function in bronchial asthma. Eur J Clin Pharmacol 3:123–125

Holmes R, Torrance RW (1959) Afferent fibres of the stellate ganglion. Q J Exp Physiol 44:271–281

Hopp FA, Zuperku EJ, Coon RL, Kampine JP (1980) Effect of anodal blockade of myelinated fibers on vagal C-fiber afferents. Am J Physiol 239:R454–R462

Hung KS, Hertweck MS, Hardy JD, Loosli CG (1972) Innervation of pulmonary alveoli of the mouse lung: an electron microscopic study. Am J Anat 135:477–496

Hung KS, Hertweck MS, Hardy JD, Loosli CG (1973a) Electron microscopic observations of nerve endings in the alveolar walls of mouse lungs. Am Rev Respir Dis 108: 328–333

Hung KS, Hertweck MS, Hardy JD, Loosli CG (1973b) Ultrastructure of nerves and associated cells in bronchiolar epithelium of the mouse lung. J Ultrastruct Res 43: 426–437

Iggo A (1958) The electrophysiological identification of single nerve fibres, with particular reference to the slowest-conducting vagal afferent fibres in the cat. J Physiol (Lond) 142:110–126

Jain SK, Subramanian S, Julka DB, Guz A (1972) Search for evidence of lung chemoreflexes in man: study of respiratory and circulatory effects of phenyldiguanide and lobeline. Clin Sci 42:163–177

Jain SK, Trenchard D, Reynolds F, Noble MIM, Guz A (1973) The effect of local anaesthesia of the airway on respiratory reflexes in the rabbit. Clin Sci 44:519–538

Jammes Y, Mei N (1979) Assessment of the pulmonary origin of bronchoconstrictor vagal tone. J Physiol (Lond) 291:305–316

Jancso G, Kiraly E, Jancso-Gabor A (1977) Pharmacologically induced selective degeneration of chemosensitive primary sensory neurons. Nature 270:741–743

Jessell TM, Iversen LL, Cuello AC (1978) Capsaicin-induced depletion of substance P from primary sensory neurones. Brain Res 152:183–188

Junod AF (1975) Metabolism, production and release of hormones and mediators in the lung. Am Rev Respir Dis 112:93–108

Junod AF (1977) Metabolism of vasoactive agents in lung. Am Rev Respir Dis 115: 51–57

Kalia M (1973) Effects of certain cerebral lesions on the J reflex. Pfluegers Arch 343: 297–308

Kalia M, Koepchen HP, Paintal AS (1973) Somatomotor and autonomous effects of type J-receptor stimulation in awake-freely moving and restrained cats. Pfluegers Arch 339: Suppl R80

Karczewski W, Widdicombe JG (1969a) The effect of vagotomy, vagal cooling and efferent vagal stimulation on breathing and lung mechanics of rabbits. J Physiol (Lond) 201:259–270

Karczewski W, Widdicombe JG (1969b) The role of the vagus nerves in the respiratory and circulatory reactions to anaphylaxis in rabbits. J Physiol (Lond) 201:293 –304

Karczewski W, Widdicombe JG (1969c) The role of the vagus nerves in the respiratory and circulatory responses to intravenous histamine and phenyl diguanide in rabbits. J Physiol (Lond) 201:271 –291

Kardon MB, Peterson DF, Bishop VS (1973) Reflex bradycardia due to aortic nerve stimulation in the rabbit. Am J Physiol 225:7 –11

Kaufman MP, Baker DG, Coleridge HM, Coleridge JCG (1980a) Stimulation by brady-kinin of afferent vagal C-fibers with chemosensitive endings in the heart and aorta of the dog. Circ Res 46:476 –484

Kaufman MP, Coleridge HM, Coleridge JCG, Baker DG (1980b) Bradykinin stimulates afferent vagal C-fibers in intrapulmonary airways of dogs. J Appl Physiol 48:511 –517

Kaufman MP, Coleridge HM, Coleridge JCG, Baker DG (1980c) Differential sensitivity of bronchial and pulmonary C-fibers in dogs to bradykinin, histamine and sero-tonin. Fed Proc 39:828

Kaufman MP, Iwamoto GA, Ashton JH, Cassidy SS (1982a) Responses to inflation of vagal afferents with endings in the lungs of dogs. Circ Res 51:525 –531

Kaufman MP, Ordway GA, Longhurst JC, Mitchell JH (1982b) Reflex relaxation of tracheal smooth muscle by thin-fiber muscle afferents in dogs. Am J Physiol 243: R383 –R388

Kawakami Y, Uchiyama K, Irie T, Murao M (1973) Evaluation of aerosols of prosta-glandins E_1 and E_2 as bronchodilators. Eur J Clin Pharmacol 6:127 –132

Keele CA, Armstrong D (1964) Substances producing pain and itch. Arnold, London

Kessler GF, Austin JHM, Graf PD, Gamsu G, Gold WM (1973) Airway constriction in experimental asthma in dogs: tantalum bronchographic studies. J Appl Physiol 35: 703 –709

Knowlton GC, Larrabee MG (1946) A unitary analysis of pulmonary volume receptors. Am J Physiol 147:100 –114

Koepchen HP, Kalia M, Sommer D, Klussendorf D (1977) Action of type J afferents on the discharge pattern of medullary respiratory neurons. In: Paintal AS, Gill-Kumar P (eds) Krogh centenary symposium on respiratory adaptations, capillary exchange and reflex mechanisms. Vallabhbhai Patel Chest Institute, Delhi, pp 407 –425

Koller EA, Ferrer P (1973) Discharge patterns of the lung stretch receptors and activation of deflation fibres in anaphylactic bronchial asthma. Respir Physiol 17: 113 –126

Kostreva DR, Zuperku EJ, Hess GL, Coon RL, Kampine JP (1975) Pulmonary afferent activity recorded from sympathetic nerves. J Appl Physiol 39:37 –40

Lahari S, Mulligan E, Nishino T, Mokashi A (1979) Aortic body chemoreceptor re-sponses to changes in Pco_2 and Po_2 in the cat. J Appl Physiol 47:858 –866

Larsell O (1921) Nerve terminations in the lung of the rabbit. J Comp Neurol 33: 105 –131

Lauweryns JM, Cokelaere M (1973) Hypoxia-sensitive neuro-epithelial bodies intra-pulmonary secretory neuroceptors, modulated by the CNS. Z Zellforsch 145: 521 –540

Lauweryns JM, Goddeeris P (1975) Neuroepithelial bodies in the human child and adult lung. Am Rev Respir Dis 111:469 –476

Lecomte J, Petit JM, Melon J, Troquet J, Marcelle R (1962) Proprietes broncho-constrictrices de la bradykinine chez l'homme asthmatique. Arch Int Pharmacodyn Ther 137:232 –235

Lee LY, Dumont C, Djokic TD, Menzel TE, Nadel JA (1979) Mechanism of rapid, shallow breathing after ozone exposure in conscious dogs. J Appl Physiol 46: 1108 –1114

Lee LY, Morton RF, Hord AH, Frazier DT (1983) Reflex control of breathing follow-
    ing inhalation of cigarette smoke in conscious dogs. J Appl Physiol 54:562–570

Leff AR, Munoz NM, Alderman B (1982) Measurement of airway response by iso-
    metric and nonisometric techniques in situ. J Appl Physiol 52:1363–1367

Levine BW, Talamo RC, Kazemi H (1973) Action and metabolism of bradykinin in
    dog lung. J Appl Physiol 34:821–826

Linden RJ, Mary DASG, Weatherill D (1981) The effect of cooling on transmission
    of impulses in vagal nerve fibres attached to atrial receptors in the dog. Q J Exp
    Physiol 66:321–332

Lloyd TC Jr (1978) Reflex effects of lung inflation and inhalation of halothane, ether
    and ammonia. J Appl Physiol 45:212–218

Lloyd TC Jr (1979) Effects of extrapulmonary airway distension on breathing in
    anesthetized dogs. J Appl Physiol 46:890–896

Loofbourrow GN, Wood WB, Baird IL (1957) Tracheal constriction in the dog. Am J
    Physiol 191:411–415

Lucas EA, Ginzel KH (1980) The effect of "J receptor" stimulation on respiration in
    the non-anesthetized cat. Fed Proc 39:1017

Lundberg JM, Saria A (1982) Capsaicin-sensitive vagal neurons involved in control of
    vascular permeability in rat trachea. Acta Physiol Scand 115:521–523

Lynn B (1977) Cutaneous hyperalgesia. Br Med Bull 33:103–108

Marshall R, Widdicombe JG (1958) The activity of pulmonary stretch receptors dur-
    ing congestion of the lungs. Q J Exp Physiol 43:320–330

Mei N (1966) Existence d'une séparation anatomique des fibres vagales efférentes et
    afférentes au niveau du ganglion plexiforme du chat. Etude morphologique et
    électrophysiologique. J Physiol (Paris) 58:253–254

Meyrick B, Reid L (1971) Nerves in rat intra-acinar alveoli: an electron microscopic
    study. Respir Physiol 11:367–377

Mills JE, Sellick H, Widdicombe JG (1969) Activity of lung irritant receptors in pul-
    monary micro-embolism, anaphylaxis and drug-induced bronchoconstrictions.
    J Physiol (Lond) 203:337–357

Mills JE, Sellick H, Widdicombe JG (1970) Epithelial irritant receptors in the lungs.
    In: Porter R (ed) Breathing: Hering-Breuer centenary symposium. Churchill,
    London, pp 77–92

Miserocchi G, Sant'Ambrogio G (1974) Distribution of pulmonary stretch receptors in
    the intrapulmonary airways of the dog. Respir Physiol 21:71–75

Miserocchi G, Trippenbach T, Mazzarelli M, Jaspar N, Hazucha M (1978) The mech-
    anism of rapid shallow breathing due to histamine and phenyldiguanide in cats and
    rabbits. Respir Physiol 32:141–153

Moncada S, Ferreira SH, Vane JR (1978) Pain and inflammatory mediators. In: Vane
    JR, Ferreira SH (eds) Inflammation. Springer, Berlin Heidelberg New York, pp
    588–616 (Handbook of experimental pharmacology, vol 50/1)

Mortola J, Sant'Ambrogio G, Clement MG (1975) Localization of irritant receptors
    in the airways of the dog. Respir Physiol 24:107–114

Morton DR, Klassen KP, Curtis GM (1951) The clinical physiology of the human
    bronchi. II. The effect of vagus section upon pain of tracheobronchial origin.
    Surgery 30:800–809

Munger BL (1971) Patterns of organization of peripheral sensory receptors. In:
    Lowenstein WR (ed) Principles of receptor physiology. Springer, Berlin Heidelberg
    New York, pp 523–556 (Handbook of sensory physiology, vol 1)

Nadel JA (1976) Airways: autonomic regulation and airway responsiveness. In: Weiss
    EB, Segal MS (eds) Bronchial asthma. Little Brown, Boston, pp 155–162

Nadel JA (1980) Autonomic regulation of airway smooth muscle. In: Nadel JA (ed)
    Physiology and pharmacology of the airways. Dekker, New York, pp 217–257

Nadel JA, Davis B (1980) Parasympathetic and sympathetic regulation of secretion
    from submucosal glands in airways. Fed Proc 39:3075–3079

Nadel JA, Widdicombe JG (1962a) Effect of changes in blood gas tensions and carotid sinus pressure on tracheal volume and total lung resistance to airflow. J Physiol (Lond) 163:13–33

Nadel JA, Widdicombe JG (1962b) Reflex effects of upper airway irritation on total lung resistance and blood pressure. J Appl Physiol 17:861–865

Nadel JA, Wolfe WG, Graf PD (1968) Powdered tantalum as a medium for bronchography in canine and human lungs. Invest Radiol 3:229–238

Nadel JA, Davis B, Phipps RJ (1979) Control of mucus secretion and ion transport in airways. Ann Rev Physiol 41:369–381

Nakano J, Rodgers RL (1976) The prostaglandins: biochemistry, pathophysiology, and clinical pharmacology in asthma and other lung disorders. In: Weiss EB, Segal MS (eds) Bronchial asthma mechanism and therapeutics. Little Brown, Boston, pp 191–216

Nathan PW, Sears TA (1961) Some factors concerned in differential nerve block by local anaesthetics. J Physiol (Lond) 157:565–580

Noble MIM, Eisele JH, Trenchard D, Guz A (1970) Effect of selective peripheral nerve blocks on respiratory sensations. In: Porter R (ed) Breathing: Hering-Breuer centenary symposium. Churchill, London, pp 233–246

Orehek J (1981) Neurohumoral control of airway caliber. In: Widdicombe JG (ed) Respiratory physiology III. University Park Press, Baltimore, pp 1–74 (International review of physiology, vol 23)

Pack AI (1981) Sensory inputs to the medulla. Annu Rev Physiol 43:73–90

Pack AI, Delaney RG, Fishman AP (1981) Augmentation of phrenic neural activity by increased rates of lung inflation. J Appl Physiol 50:149–161

Paintal AS (1955) Impulses in vagal afferent fibres from specific pulmonary deflation receptors. The response of these receptors to phenyl diguanide, potato starch, 5-hydroxytryptamine and nicotine, and their role in respiratory and cardiovascular reflexes. Q J Exp Physiol 40:89–111

Paintal AS (1957) The location and excitation of pulmonary deflation receptors by chemical substances. Q J Exp Physiol 42:56–71

Paintal AS (1963) Vagal afferent fibres. Ergeb Physiol 52:74–156

Paintal AS (1964) Effects of drugs on vertebrate mechanoreceptors. Pharmacol Rev 16:341–380

Paintal AS (1965) Block of conduction in mammalian myelinated nerve fibres by low temperatures. J Physiol (Lond) 180:1–19

Paintal AS (1966) Re-evaluation of respiratory reflexes. Q J Exp Physiol 51:151–163

Paintal AS (1967) Mechanism of stimulation of aortic chemoreceptors by natural stimuli and chemical substances. J Physiol (Lond) 189:63–84

Paintal AS (1969) Mechanism of stimulation of type J pulmonary receptors. J Physiol (Lond) 203:511–532

Paintal AS (1970) The mechanism of excitation of type J receptors, and the J reflex. In: Porter R (ed) Breathing: Hering-Breuer centenary symposium. Churchill, London, pp 59–71

Paintal AS (1973) Vagal sensory receptors and their reflex effects. Physiol Rev 53: 159–227

Paintal AS (1974) Fluid pump of type J receptors of the cat. J Physiol (Lond) 238: 53–54P

Paintal AS (1977a) The nature and effects of sensory inputs into the respiratory centers. Fed Proc 36:2428–2432

Paintal AS (1977b) Thoracic receptors connected with sensation. Br Med Bull 33: 169–174

Paintal AS, Damodaran VN, Guz A (1973) Mechanism of excitation of type J receptors. Acta Neurobiol Exp 33:15–19

Phillipson EA, Hickey RF, Bainton CR, Nadel JA (1970) Effect of vagal blockade on regulation of breathing in conscious dogs. J Appl Physiol 29:475–479

Phillipson EA, Hickey RF, Graf PD, Nadel JA (1971) Hering-Breuer inflation reflex and regulation of breathing in conscious dogs. J Appl Physiol 31:746–750

Phillipson EA, Fishman NH, Hickey RF, Nadel JA (1973) Effect of differential vagal blockade on ventilatory response to CO_2 in awake dogs. J Appl Physiol 34:759–763

Phillipson EA, Murphy E, Kozar LF (1975a) Role of pulmonary J receptors in regulation of ventilation in conscious dogs. Fed Proc 34:371

Phillipson EA, Murphy E, Kozar LF, Schultze RK (1975b) Role of vagal stimuli in exercise ventilation in dogs with experimental pneumonitis. J Appl Physiol 39:76–85

Phipps RJ, Richardson PS (1976) The effects of irritation at various levels of the airway upon tracheal mucus secretion in the cat. J Physiol (Lond) 261:563–581

Piper P, Vane J (1971) The release of prostaglandins from lung and other tissues. Ann NY Acad Sci 180:363–385

Plaut M, Lichtenstein LM (1978) Histamine, 5-hydroxytryptamine, SRS-A: Discussion of type I hypersensitivity (anaphylaxis). In: Vane JR, Ferreira SH (eds) Inflammation. Springer, Berlin Heidelberg New York, pp 345–373 (Handbook of experimental pharmacology, vol 50/1)

Porszasz J, Such G, Porszasz-Gibiszer K (1957) Circulatory and respiratory chemoreflexes. I. Analysis of the site of action and receptor types of capsaicine. Acta Physiol Acad Sci Hung 12:189–205

Rao KS, Devanandan MS (1977) Spinal organisation of the reflex inhibition of the skeleto-motor system by activation of type J pulmonary afferents. In: Paintal AS, Gill-Kumar P (eds) Krogh centenary symposium on respiratory adaptations, capillary exchange and reflex mechanisms. Vallabhabhai Patel Chest Institute, Delhi, pp 466–484

Raybould HE, Russell NJW (1982) Afferent activity in pulmonary vagal C-fibres reflexly increases respiratory rate during hypercapnia in the anaesthetized rabbit. J Physiol (Lond) 326:60–61P

Reid L (1960) Chronic bronchitis and hypersecretion of mucus. Lect Sci Basis Med 8:235–255

Reynolds LB Jr (1962) Characteristics of an inspiration-augmenting reflex in anesthetized cats. J Appl Physiol 17:683–688

Rhodin JAG (1966) Ultrastructure and function of the human tracheal mucosa. Am Rev Respir Dis 93:1–15

Richardson PS, Widdicombe JG (1969) The role of the vagus nerves in the ventilatory responses to hypercapnia and hypoxia in anaesthetized and unanaesthetized rabbits. Respir Physiol 7:122–135

Richardson CA, Herbert DA, Mitchell RA (1982) Pulmonary resistance is modulated in step with inspiratory output in paralyzed cats. Physiologist 25:189

Roberts AM, Armstrong DJ, Coleridge HM, Coleridge JCG (1981a) Paradoxical tracheal contraction produced by bronchodilator prostaglandins PGE_2 and PGI_2. Fed Proc 40:595

Roberts AM, Kaufman MP, Baker DG, Brown JK, Coleridge HM, Coleridge JCG (1981b) Reflex tracheal contraction induced by stimulation of bronchial C-fibers in dogs. J Appl Physiol 51:485–493

Roberts AM, Coleridge HM, Coleridge JCG (1982a) Reciprocal action of pulmonary stretch receptors and lung C-fibers on tracheal smooth muscle tone in dogs. Fed Proc 41:986

Roberts AM, Hahn HL, Schultz HD, Nadel JA, Coleridge HM, Coleridge JCG (1982b) Afferent vagal C fibers are responsible for the reflex airway constriction and secretion evoked by pulmonary administration of SO_2 in dogs. Physiologist 25:226

Roumy M, Leitner LM (1980) Localization of stretch and deflation receptors in the airways of the rabbit. J Physiol (Paris) 76:67–70

Russell JA (1978) Responses of isolated canine airways to electrical stimulation and acetylcholine. J Appl Physiol 45:690–698

Russell JA, Lai-Fook SJ (1979) Reflex bronchoconstriction induced by capsaicin in the dog. J Appl Physiol 47:961–967

Russell NJW, Trenchard D (1980) Non-myelinated vagal lung receptors in the rabbit. J Physiol (Lond) 300:31P

Russell NJW, Trenchard D (1981) Chemoreflexes of pulmonary origin elicited by sodium dithionite in the anaesthetized rabbit. J Physiol (Lond) 310:63–64P

Salisbury PF, Galletti PM, Lewin RJ, Rieben PA (1959) Stretch reflexes from the dog's lung to the systemic circulation. Circ Res 7:62–67

Sampson SR (1977) Sensory neurophysiology of airways. Am Rev Respir Dis 115:107–115

Sampson SR, Vidruk EH (1975) Properties of 'irritant' receptors in canine lung. Respir Physiol 25:9–22

Sampson SR, Vidruck EH (1978) Chemical stimulation of rapidly adapting receptors in the airways. In: Fitzgerald RS, Gautier H, Lahiri S (eds) The regulation of respiration during sleep and anesthesia. Adv Exp Med Biol 99:281–290

Sant'Ambrogio FB, Sant'Ambrogio G (1982) Circulatory accessibility of nervous receptors localized in the tracheobronchial tree. Respir Physiol 49:49–73

Sant'Ambrogio G (1982) Information arising from the tracheobronchial tree of mammals. Physiol Rev 62:531–569

Sant'Ambrogio G, Sant'Ambrogio FB, Davies A (1980) Coughing to irritation of the tracheo-bronchial tree and the larynx after sulphur dioxide inhalation in rabbits in dogs. Physiologist 23:88

Sapru HN, Willette RN, Krieger AJ (1981) Stimulation of pulmonary J-receptors by an enkephalin-analog. J Pharmacol Exp Ther 217:228–234

Satchell GH (1977) The J reflex in fish. In: Paintal AS, Gill-Kumar P (eds) Krogh centenary symposium on respiratory adaptations, capillary exchange and reflex mechanisms. Vallabhbhai Patel Chest Institute, Delhi, pp 432–439

Schmidt CF (1941) The reflex regulation of respiration. In: Bard P (ed) MacLeod's physiology in modern medicine, 9th edn. Kimpton, London, pp 569–600

Schmidt T, Wellhoner HH (1970) The reflex influence of a group of slowly conducting vagal afferents on α and γ discharges in cat intercostal nerves. Pfluegers Arch 318:335–345

Schulte FJ, Henatsch HD, Busch G (1959) Über den Einfluß der Carotid-sinus-Sensibilität auf die spinalmotorischen Systeme. Pfluegers Arch 269:248–263

Schweitzer A, Wright S (1937) Effects on the knee jerk of stimulation of the central end of the vagus and of various changes in the circulation and respiration. J Physiol (Lond) 88:459–475

Sellick H, Widdicombe JG (1969) The activity of lung irritant receptors during pneumothorax, hyperpnoea and pulmonary vascular congestion. J Physiol (Lond) 203:359–381

Sellick H, Widdicombe JG (1970) Vagal deflation and inflation reflexes mediated by lung irritant receptors. Q J Exp Physiol 55:153–163

Sellick H, Widdicombe JG (1971) Stimulation of lung irritant receptors by cigarette smoke, carbon dust, and histamine aerosol. J Appl Physiol 31:15–19

Shanes AM, Gershfeld NL (1960) Interactions of veratrum alkaloids, procaine, and calcium with monolayers of stearic acid and their implications for pharmacological action. J Gen Physiol 44:345–363

Sheldon MI, Green JF (1982) Evidence for pulmonary CO_2 chemosensitivity: effects on ventilation. J Appl Physiol 52:1192–1197

Simonsson BG, Skoogh BE, Bergh NP, Andersson R, Svedmyr N (1973) In vivo and in vitro effect of bradykinin on bronchial motor tone in normal subjects and patients with airways obstruction. Respiration 30:378–388

Smith AP (1973) The effects of intravenous infusion of graded doses of prostaglandins $F_{2\alpha}$ and E_2 on lung resistance in patients undergoing termination of pregnancy. Clin Sci 44:17–25

Smith AP, Cuthbert MF (1976) The response of normal and asthmatic subjects to prostaglandins E_2 and $F_{2\alpha}$ by different routes, and their significance in asthma. In: Samuelsson B, Paoletti R (eds) Advances in prostaglandin and thromboxane research, vol 1. Raven Press, New York, pp 449–459

Souhrada JF, Dickey DW (1976) Mechanical activities of trachea as measured in vitro and in vivo. Respir Physiol 26:27–40

Stephens NL, Kroeger EA (1980) Ultrastructure, biophysics, and biochemistry of airway smooth muscle. In: Nadel JA (ed) Physiology and pharmacology of the airways. Dekker, New York, pp 31–121

Stern S, Bruderman I, Braun K (1966) Localization of lobeline-sensitive receptors in the pulmonary circulation in man. Am Heart J 71:651–655

Stransky A, Szereda-Przestaszewska M, Widdicombe JG (1973) The effects of lung reflexes on laryngeal resistance and motoneurone discharge. J Physiol (Lond) 231:417–438

Thorén P, Mancia G, Shepherd JT (1975) Vasomotor inhibition in rabbits by vagal non-medullated fibers from cardiopulmonary area. Am J Physiol 229:1410–1413

Thorén P, Shepherd JT, Donald DE (1977) Anodal block of medullated cardiopulmonary vagal afferents in cats. J Appl Physiol 42:461–465

Toh CC, Lee TS, Kiang AK (1955) The pharmacological actions of capsaicin and analogues. Br J Pharmacol 10:175–182

Tomori Z, Widdicombe JG (1969) Muscular bronchomotor and cardiovascular reflexes elicited by mechanical stimulation of the respiratory tract. J Physiol (Lond) 200: 25–49

Torebjörk HE, Hallin RG (1974) Identification of afferent C units in intact human skin nerves. Brain Res 67:387–403

Trenchard D (1977) Role of pulmonary stretch receptors during breathing in rabbits, cats and dogs. Respir Physiol 29:231–246

Trenchard D (1980) A reappraisal of vagal lung chemoreflex changes in breathing. J Physiol (Lond) 302:16–17P

Trenchard D, Gardner D, Guz A (1972) Role of pulmonary vagal afferent nerve fibres in the development of rapid shallow breathing in lung inflammation. Clin Sci 42: 251–263

Waaler BA (1961) The effect of bradykinin in an isolated perfused dog lung preparation. J Physiol (Lond) 157:475–483

Wasserman MA (1975) Bronchopulmonary responses to prostaglandin $F_{2\alpha}$, histamine and acetylcholine in the dog. Eur J Pharmacol 32:146–155

Whitteridge D (1948) The action of phosgene on the stretch receptors of the lungs. J Physiol (Lond) 107:107–114

Whitteridge D (1950) Multiple embolism of the lung and rapid shallow breathing. Physiol Rev 30:475–486

Whitteridge D, Bulbring E (1944) Changes in activity of pulmonary receptors in anaesthesia and their influence on respiratory behavior. J Pharmacol Exp Ther 81: 340–359

Whitwam JG, Kidd C (1975) The use of direct current to cause selective block of large fibres in peripheral nerves. Br J Anaesth 47:1123–1132

Widdicombe JG (1954a) Receptors in the trachea and bronchi of the cat. J Physiol (Lond) 123:71–104

Widdicombe JG (1954b) Respiratory reflexes excited by inflation of the lungs. J Physiol (Lond)123:105–115

Widdicombe JG (1954c) Respiratory reflexes from the trachea and bronchi of the cat. J Physiol (Lond) 123:55–70

Widdicombe JG (1963) Regulation of tracheobronchial smooth muscle. Physiol Rev 43:1–37

Widdicombe JG (1964) Respiratory reflexes. In: Fenn WO, Rahn H (eds) Respiration. American Physiological Society, Washington, DC, pp 585–630 (Handbook of physiology, vol 1)

Widdicombe JG (1966) Action potentials in parasympathetic and sympathetic efferent fibres to the trachea and lungs of dogs and cats. J Physiol (Lond) 186:56–88

Widdicombe JG (1967) Head's paradoxical reflex. Q J Exp Physiol 52:44–50

Widdicombe JG (1974a) Reflex control of breathing. In: Widdicombe JG (ed) Respiratory physiology. University Park Press, Baltimore, pp 273–301 (MTP International review of science, ser 1, Physiology, vol 2)

Widdicombe JG (1974b) Reflexes from the lungs in the control of breathing. In: Linden RJ (ed) Recent advances in physiology. Churchill, London, pp 239–278

Widdicombe JG (1977a) Defensive mechanisms of the respiratory system. In: Widdicombe JG (ed) International review of physiology, vol 14. Respiratory physiology II. University Park Press, Baltimore, pp 291–315

Widdicombe JG (1977b) Some experimental models of acute asthma. J R Coll Physicians Lond 11:141–155

Widdicombe JG (1978) Control of secretion of tracheobronchial mucus. Br Med Bull 34:57–61

Widdicombe JG (1981) Nervous receptors in the respiratory tract and lungs. In: Hornbein TH (ed) Regulation of breathing, part I. Dekker, New York, pp 429–472

Widdicombe JG, Nadel JA (1963) Reflex effects of lung inflation on tracheal volume. J Appl Physiol 18:681–686

Winning AJ, Widdicombe JG (1976) The effect of lung reflexes on the pattern of breathing in cats. Respir Physiol 27:253–266

Woolcock AJ, Macklem PT, Hogg JC, Wilson NJ, Nadel JA, Frank NR, Brain J (1969) Effect of vagal stimulation on central and peripheral airways in dogs. J Appl Physiol 26:806–813

Yamatake Y, Yanaura S (1978) New method for evaluating bronchomotor and bronchosecretory activities: effects of prostaglandins and antigen. Jpn J Pharmacol 28:391–402

Rev. Physiol. Bio chem. Pharmacol., Vol. 99
© by Springer-Verlag 1984

Biology and Biochemistry of Papillomaviruses

HERBERT PFISTER *

Contents

* Institut für Klinische Virologie, Loschgestraße 7, 8520 Erlangen, FRG

1 Introduction

Papillomaviruses induce benign tumors in skin and mucosa. In the case of human warts the viral etiology was suggested by transmission with cell-free filtrates as early as 1907 (*Ciuffo* 1907). In contrast to most other DNA tumor viruses, infectious papillomavirus particles are synthesized within the tumor tissue. They can be disclosed in ultrathin sections by the electron microscope, extracted from ground biopsies, purified, and used for reinfection experiments, thus fulfilling Koch's postulates to link an infectious agent and a disease. Virus-induced papillomas of humans and animals may undergo malignant transformation to carcinomas (*Zur Hausen* 1977a). Viral DNA was shown to persist in some of the malignant tumors, and there is growing interest in the role of the virus during malignant conversion.

Papillomaviruses are classified as genus *Papillomavirus* of the family Papovaviridae (*Matthews* 1982). Polyoma virus and simian virus 40, among others, form the second genus. Both genera were grouped together on the basis of similar structures of capsid and nucleic acid. They are distinguished by the size of the capsid (55 nm vs 40 nm) and by the molecular weight of the nucleic acid (5×10^6 vs 3.3×10^6). Whereas the molecular biology of polyoma-like tumor viruses has been investigated in great detail, papillomavirus research has been mainly hampered by the lack of suitable cell culture systems for in vitro propagation. Recent technical advances, however, especially the molecular cloning of DNA, have led to a renaissance of papillomaviruses and to a great expansion of our knowledge of their biology and biochemistry. It is now realized that they are quite unrelated to polyoma-like viruses and are likely to reveal a number of unique properties.

It is the aim of this review to summarize recent data on the molecular biology of the papillomavirus group as a whole and to discuss them in the light of earlier reports on biology and pathology. The human papillomaviruses will be treated in more detail as far as clinical features, immunology, diagnosis, and treatment are concerned. The literature cited on malignant conversion of human papillomas is not complete, and the reader should refer to the comprehensive review of *Zur Hausen* (1977a). With regard to the special aspects of animal papillomavirus biology and immunology, the reader is referred to the work of *Lancaster* and *Olson* (1982).

2 Abbreviations

A	Adenine
bp	base pairs = pairs of nucleotides
BPV	bovine papillomavirus
C	cytosine
COPV	canine oral papillomavirus
CRPV	cottontail rabbit (Shope) papillomavirus
DNA	deoxyribonucleic acid
Ev	epidermodysplasia verruciformis
G	guanine
HPV	human papillomavirus
MnPV	*Mastomys natalensis* papillomavirus (virus of the multi-mammate mouse)
mRNA	messenger ribonucleic acid
RNA	ribonucleic acid
SV40	simian virus 40
T	thymine

3 Properties of the Virions

Papillomavirus particles are about 55 nm in diameter according to electron microscopic measurements and have an icosahedral symmetry with
72 capsomers (*Strauss* et al. 1950; *Finch* and *Klug* 1965; *Klug* and *Finch*
1965). After negative staining the capsomers appear as hollow cylinders of
equal height and width, which are connected at their base by fibrous
bridge-like structures (*Yabe* et al. 1979). The virions consist only of DNA
and protein. Complete particles show a density of 1.34 g/cm³ in CsCl and
are easily separated by equilibrium gradient centrifugation from "empty"
capsids without DNA (density: 1.29 g/cm³) (*Breedis* et al. 1962). Tubular
capsids may be detected in the band of lower density (*Noyes* 1965). The
tubes are made up of hexagonal capsomers and their diameter is close to
that of the virions. The filaments are assumed to result from aberrant
maturation.

Papillomavirus DNA is a double-stranded, circular molecule and exists
in two major forms: (1) with both strands covalently closed, which leads
to a rigid topological constraint and to superhelical twists; and (2) with
one single-strand nick, which leads to the relaxed or open-circle form. The
linear form, resulting from one double-strand break, is rarely detected in
gently isolated material.

The DNAs of most virus types have an average molecular weight of
5.0×10^6, corresponding to approximately 8000 bp (*Kleinschmidt* et al.
1965; *Orth* et al. 1977, 1978b; *Gissmann* and *Zur Hausen* 1980; *Chen* et
al. 1982; *Danos* et al. 1982). The DNAs of some papillomaviruses deviate
considerably: BPV 3 and BPV 4 are dwarfs, having DNAs with a molecular
weight of 4.4×10^6 (*Pfister* et al. 1979c; *Campo* et al. 1980); whereas the
cutaneous virus from the European elk (*Moreno Lopez* et al. 1981) and
the COPV of dogs (*Pfister* and *Meszaros* 1980) are giants, the molecular
weight of their DNAs being about 5.5×10^6. The G + C content varies
between 41% (HPV 1) and 50% (MnPV) (*Crawford* and *Crawford* 1963;
Müller and *Gissmann* 1978). Besides type-specific size differences, *Yoshiike*
and *Defendi* (1977) observed deletions in HPV DNA which was derived
from pooled warts. The extent of deletions ranged from 7% to 24% of the
entire genome.

In BPV 1 and 2, the DNA was shown to be associated with histone-
related proteins to form a chromatin-like complex (*Favre* et al. 1977).
Similar proteins of HPV 1 comigrate with cellular histones H2a, H2b, H3,
and H4 in sodium dodecyl sulfate (SDS) polyacrylamide gel electrophoresis. Differences were noted in an acetic acid/urea system for the H3- and
H4-like proteins (*Pfister* and *Zur Hausen* 1978a). These differences might
be due to acetylation, as described for polyoma virus (*Schaffhausen* and

Benjamin 1976). Proteins, which comigrate with cellular histones, were not detected in preparations of HPV 4 or BPV 3 (*Gissmann* et al. 1977; *Pfister* et al. 1979c).

Many papillomavirus types are only produced in very small amounts, which do not permit protein analysis. Structural proteins could be studied with HPV 1, HPV 2, HPV 4, BPV 1–3, CRPV, and MnPV (*Favre* et al. 1975a; *Gissmann* et al. 1977; *Orth* et al. 1977; *Lancaster* and *Olson* 1978; *Müller* and *Gissmann* 1978; *Pfister* et al. 1979c). The molecular weights of the major structural proteins were in the range of 53 000–59 000. Amino acid analysis of the BPV 1 protein revealed a highly acidic protein containing almost twice the average number of acidic residues as compared to basic residues (*Meinke* and *Meinke* 1981). In HPV 1, additional proteins with molecular weights of 53 000 and 43 000, respectively, were described. Analysis of BrCN peptides indicated that they may result from conversion of the major polypeptide (*Pfister* et al. 1977). The virus specificity and location of a number of minor protein components remains to the established.

3.1 Human Papillomaviruses

The number of HPV types has almost exponentially increased during the past few years. Many of them could not have been characterized directly from wart material as they are produced in very low quantity. Due to the absence of a tissue culture system for the propagation of papillomaviruses, analysis only became feasible after molecular cloning either in bacterial plasmids or in bacteriophage lambda (*Heilman* et al. 1980; *De Villiers* et al. 1981; *Pfister* et al. 1981b, 1983b; *Gissmann* et al. 1982b; *Kremsdorf* et al. 1982). At present a new virus isolate is considered an independent type if nucleic acid hybridization studies reveal less than 50% homology with known virus types (*Coggin* and *Zur Hausen* 1979). Isolates which show more than 50% cross-hybridization but differ in their restriction endonuclease cleavage patterns are regarded as subtypes. The classification may be supported by serological data, but in most cases this will be difficult or impossible because of limiting amounts of antigen. The border line of 50% cross-hybridization is certainly arbitrary. However, since the field is rapidly evolving at the moment and there is still little information on the biological properties of some virus types, the present agreement is likely to provide the most reasonable classification scheme to avoid nomenclature confusion.

On the basis of less than 50% cross-hybridization, 16 HPV types have been differentiated up to now. Their biological properties will be discussed in Sect. 4. Under stringent reaction conditions no hybridization was

Table 1. Human papillomavirus types

Group	Type	Reference	Reference for DNA cloning
A	HPV 1	*Favre* et al. (1975b)	*Danos* et al. (1980)
			Heilman et al. (1980)
B	HPV 2	*Orth* et al. (1977)	*Heilman* et al. (1980)
	HPV 3	*Orth* et al. (1978b)	*Gassenmaier* and *Pfister*
			(unpublished work)
	HPV 10	*Orth* (personal communication)	*Gissmann* and *Pfister*
		Green et al. (1982)	(unpublished work)
C	HPV 4	*Gissmann* et al. (1977)	*Heilman* et al. (1980)
D	HPV 5	*Orth* et al. (1980)	*Kremsdorf* et al. (1982)
	HPV 8	*Orth* et al. (1980)	*Pfister* et al. (1981b)
	HPV 9	*Orth* et al. (1980)	*Kremsdorf* et al. (1982)
	HPV 12	*Orth* (personal communication)	
	HPV 14	*Orth* (personal communication)	
	HPV 15	*Orth* (personal communication)	
E	HPV 6	*Gissmann* and *Zur Hausen* (1980)	*De Villiers* et al. (1981)
	HPV 11	*Gissmann* et al. (1982b)	*Gissmann* et al. (1982b)
	HPV 13	*Pfister* et al. (1983b)	*Pfister* et al. (1983b)
F	HPV 7	*Orth* et al. (1981)	
		Ostrow et al. (1981)	
G	HPV 16	*Dürst* et al. (1983)	*Dürst* et al. (1983)

The HPV types are arranged in seven groups, which show less than 1% DNA–DNA cross-hybridization. The members of individual groups cross-hybridize from 1% to 30%.

Table 2. Cross-hybridization between human papillomaviruses [a]

Group B [b]	HPV 2	HPV 3	HPV 10
HPV 2	100	ND	6 [c]
HPV 3		100	36 [c]
HPV 10			100

Group D [b]	HPV 5	HPV 8	HPV 9
HPV 5	100	17 [d]	5 [d]
HPV 8		100	2 [d]
HPV 9			100

Group E [b]	HPV 6	HPV 11	HPV 13
HPV 6	100	25 [e]	4 [f]
HPV 11		100	3 [f]
HPV 13			100

[a] Expressed as percentage of homologous hybridization
[b] See Table 1
[c] *Green* et al. (1982)
[d] *Kremsdorf* et al. (1982)
[e] *Gissmann* et al. (1982b)
[f] *Pfister* et al. (1983b)

detected between DNAs of HPV 1, 4, 7, and 16 (*Gissmann* et al. 1977; *Orth* et al. 1981; *Dürst* et al. 1983). The other types form three groups (Table 1), the members of which show cross-hybridization from 1% to 35% (Table 2). No significant sequence homology could be detected between members of different groups.

HPV 1, 2, 3, 4, 5, 8, and 9 were tested for serological cross-reactivity by complement fixation assays, by indirect immunofluorescence tests or by radioimmunoassays (*Gissmann* et al. 1977; *Orth* et al. 1977, 1978b; *Pfister* and *Zur Hausen* 1978b). Specific sera were obtained by immunization of rabbits or guinea pigs with defined virus isolates or with a pool containing HPV 5, 8, and 9. No cross-reaction was detected between different virus types, and the HPV 5, 8, and 9 group proved to be unrelated to the other types.

For each type of HPV subtypes probably exist, which show extensive sequence homology but differ in a number of restriction enzyme cleavage sites. This has been described in detail for HPV 1 (*Gissmann* and *Zur Hausen* 1976; *Gissmann* et al. 1977; *Pfister* 1980), HPV 2 (*Orth* et al. 1977; *Heilman* et al. 1980), and HPV 4 (*Heilman* et al. 1980; *Pfister* 1980).

Representatives of most HPV types have been molecularly cloned. The DNAs were characterized by cleavage with restriction enzymes, and physical maps of the resulting DNA fragments were established (for reference see Table 1). The DNAs of HPV 1 and HPV 6 have been completely sequenced (*Clad* et al. 1982; *Danos* et al. 1982; *Schwarz* et al., to be published). They have a length of 7812 and 7903 nucleotides, respectively.

The nucleotide sequences confirmed earlier observations and revealed further interesting features. By self-annealing of HPV 1 DNA single strands, two short palindromic sequences were detected by *Gissmann* and *Zur Hausen* (1977). As seen from the sequence they are 33 and 69 nucleotides in length and show about 30% mismatches.

A homology between HPV 1 DNA and the consensus sequence of the human Alu family of ubiquitous repeats was disclosed in the vicinity of the HPV 1–Bam HI cleavage site. One homologous region is also related to the DNA sequence at the origin of replication of SV 40, polyoma, BK virus, and hepatitis B virus.

The location of protein reading frames and of eukaryotic transcription control signals as derived from the DNA sequence will be discussed in Sects. 3.3 and 6.3.3).

3.2   Animal Papillomaviruses

Papillomaviruses are widespread among mammals. They have been described in cattle, sheep, goats, deer, elks, horses, rabbits, multimammate
mice, dogs, monkeys, pigs, opposums, and elephants (*Rangan* et al. 1980;
Sundberg et al. 1981; *Lancaster* and *Olson* 1982). There is one avian virus
infecting the chaffinch (*Osterhaus* et al. 1977).

Anomg animal papillomaviruses those from cattle have been investigated most extensively. Five types of BPV have been identified, which can
be distinguished by their biological properties. BPV 1 and 2 were repeatedly isolated from cutaneous fibropapillomas (*Lancaster* and *Olson* 1978;
Pfister et al. 1979c) and BPV 2 also from fibropapillomas of the alimentary tract (*Jarrett* 1980). BPV 3 was isolated from two cases of epithelial
proliferations on the skin (*Pfister* et al. 1979c), so-called atypical bovine
papillomas (*Koller* and *Olson* 1972). BPV 4 induces epithelial lesions of
the alimentary tract (*Campo* et al. 1980), and BPV 5 was recovered from
"rice-grain"-like papillomas of the teats (*Campo* et al. 1981).

BPV 1, 2, and 5 form one group, where BPV 1 and 2 show extensive
cross-hybridization of about 45% when tested by DNA—DNA reassociation
kinetics (*Lancaster* and *Olson* 1978); BPV 5 seems to be rather distantly
related, revealing at most 5% DNA homology with BPV 2 (*Campo* et al.
1981). In all three viruses, however, common sequences are equally distributed over the whole genome, as deduced from Southern blot hybridization between labeled probes and subgenomic restriction enzyme fragments (*Law* et al. 1979; *Pfister* et al. 1980; *Campo* et al. 1981). Furthermore, common sequences seem to be highly conserved, as shown by ther-
mal denaturation experiments with BPV 2 and BPV 5 DNAs (*Campo* et
al. 1981). BPV 1 and 2 are also antigenically cross-reactive, but titers of
monospecific animal antisera differed both in hemagglutination inhibition and complement fixation assays by a factor of 10 (*Lancaster* and
Olson 1978). The molecular basis for agglutination of mouse erythrocytes,
first noted by *Favre* et al. (1974), is unknown. BPV 5 and BPV 2 cross-
reacted only poorly in immunohistochemical tests (*Jarrett*, unpublished
work).

BPV 3 (*Pfister* et al. 1979c) and BPV 4 (*Campo* et al. 1980) form the
second group of papillomaviruses from cattle. Their DNAs do not cross-
hybridize with probes from the other group and there is no serological
cross-reactivity. The DNAs of both viruses have a molecular weight of
about 4.4×10^6, which is significantly smaller than that of BPV 1, 2, or 5
DNA (5×10^6). BPV 3 and 4 share at most 12% of their DNA sequences,
as deduced from reassocation kinetics under stringent conditions (*Pfister*,
unpublished work). As discussed for BPV 1, 2, and 5, the homologous
sequences are distributed over the whole genome, according to hetero-

duplex analysis and Southern blot hybridization with subgenomic DNA fragments (*Campo* and *Pfister*, unpublished work).

The DNAs of BPV 1−4 were molecularly cloned (*Howley* et al. 1980; *Campo* and *Coggins* 1982; *Pfister* and *Hettich*, unpublished work), and the complete nucleotide sequence of BPV 1 has been established (*Chen* et al. 1982). The DNAs of BPV 2 (*Lancaster* 1979; *Pfister* et al. 1979c; *Campo* et al. 1981; *Murphy* et al. 1981), BPV 3 (*Pfister* et al. 1979c), BPV 4 (*Campo* et al. 1980), and BPV 5 (*Campo* et al. 1981) were characterized by cleavage with restriction enzymes and the resulting DNA fragments were mapped. A comparison of physical maps reveals a close relationship between BPV 1 and BPV 2 (*Lancaster* 1979; *Murphy* et al. 1981). Hetero-duplex analysis confirmed that both DNAs are in register when aligned at their single Hind III sites (*Campo* and *Coggins* 1982). The transforming region of the BPV 1 genome (see Sect. 6.5) proved to be almost complete-ly homologous to BPV 2, whereas only partial homology was observed with the segment coding for structural proteins (see Sect. 6.3.2). This is in line with the limited serological cross-reactivity between BPV 1 and 2.

Papillomaviruses from other animal species were investigated less systematically, and only one representative was usually described on the molecular level. Data exist for viruses from rabbits (CRPV; *Favre* et al. 1975, 1982; *Murphy* et al. 1981), dogs (COPV; *Pfister* and *Meszaros* 1980), elks (*Moreno-Lopez* et al. 1981), deer (*Lancaster* and *Sundberg* 1982), multimammate mice (MnPV; *Müller* and *Gissmann* 1978), and chaffinches (*Osterhaus* et al. 1977). Comparison of DNAs from deer fibro-ma virus and BPV 1 or BPV 2 indicated a 3%−9% sequence homology (*Lancaster* and *Sundberg* 1982). This is the only example of a relationship between viruses of different species, when tested under stringent hybrid-ization conditions.

3.3 Conserved Sequences, Group-Specific Antigens

A relationship between DNAs from different papillomavirus types both of human and animal origin was revealed by hybridization under so-called relaxed conditions. Incubation at 50°C below the melting temperature of the DNA allows formation of stable hybrids between single strands show-ing up to 30% mismatch. DNAs from HPV 1, 2, 4, 6, and 8, BPV 1 and 2, CRPV, and COPV proved to be related in certain fragments when tested in this way (*Law* et al. 1979; *Pfister* and *Meszaros* 1980; *Heilman* et al. 1980; *Schulte* and *Pfister*, unpublished work). These homologous sequen-ces are not equally distributed over the genome. If relaxed hybridization is carried out with subgenomic labeled probes, it is possible to compare the genome structure of different viruses, as shown in Fig. 1. In the cases

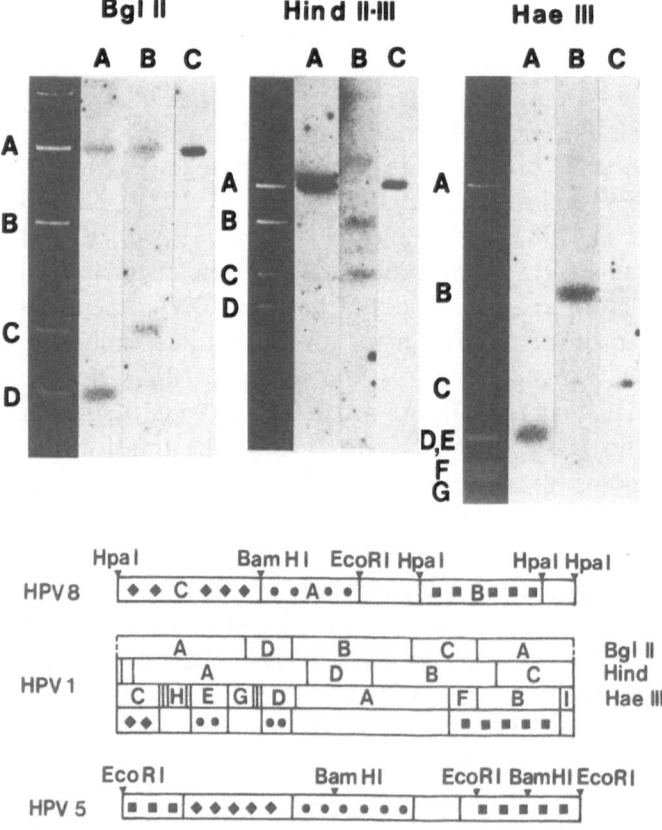

Fig. 1. Aligning of HPV genomes according to data from Southern blot hybridization with subgenomic probes under relaxed conditions (10% formamide, 1 M NaCl, 37°C). HPV 1 DNA was cleaved with the restriction enzymes Bgl II, Hind II + III, and Hae III. The fragment pattern as revealed by ethidium bromide straining is shown in the *left slot* in each case and the physical maps are given *below*. Slots *A, B,* and *C* show the result of Southern blot hybridization with ^{32}P-labeled fragments A, B, and C of HPV 8 DNA (see physical map of HPV 8). The data obtained for the three HPV 1 cleavage patterns were summarized and homologous regions between HPV 8 and HPV 1 are indicated by common symbols. The outcome of the comparison between HPV 8 and HPV 5 is demonstrated (experimental data not shown). (*Schulte* and *Pfister,* unpublished work)

of HPV 1, 5, 8, and BPV 1, the homologous regions were colinear (*Freese* et al. 1982; *Schulte* and *Pfister,* unpublished work). Congruent results were obtained by electron microscopic analysis of heteroduplex molecules (*Croissant* et al. 1982; *Croissant,* personal communication). In the latter test, HPV 1 and BPV 1 revealed three regions of about 75% homology, which represent 13% of the genome length (Fig. 2).

In Southern blot hybridization experiments some DNA segments already cross-hybridize at 36°C below the melting temperature of the

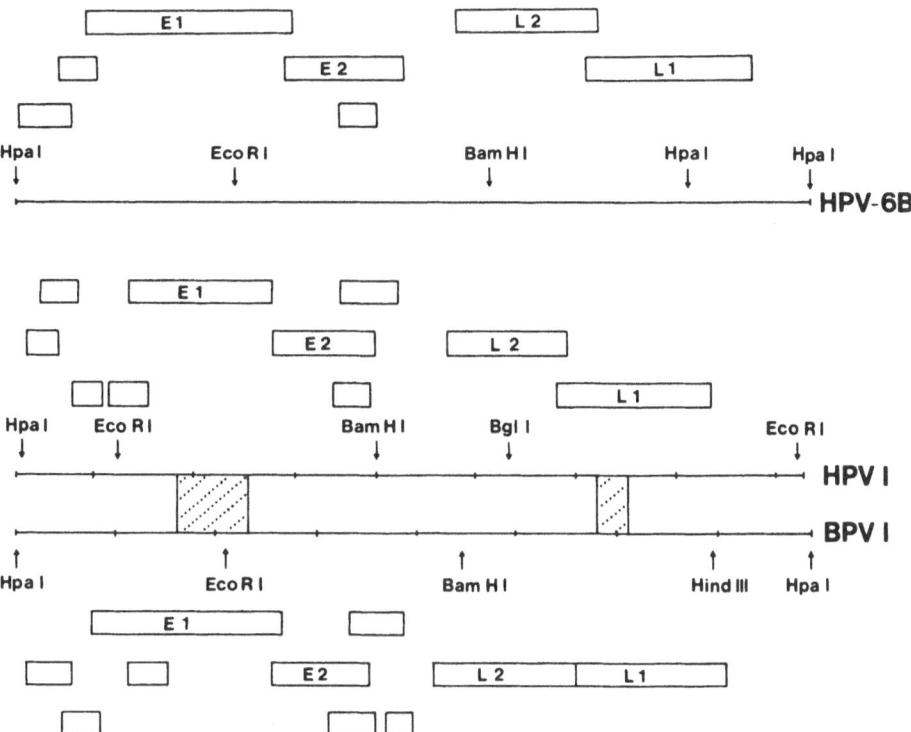

Fig. 2. Genome organization of HPV 1 (*Danos* et al. 1982), BPV 1 (*Chen* et al. 1982) and HPV 6 B (*Schwarz* et al., to be published). HPV 1 and BPV 1 were aligned according to conserved regions revealed by heteroduplex analysis (*dotted areas; Croissant* et al. 1982). Open reading frames for each potential translation frame are depicted *above* and *below* the genomic maps. The HPV 6 B genome was aligned according to sequence homologies. The two major reading frames within transforming ("early") and capsid coding ("late") genome region are labeled *E 1, E 2* and *L 1, L 2*, respectively (for explanation see Sects. 3.3 and 6.3–6.5 of the text)

DNA, indicating an even more close relationship (*Law* et al. 1979; *Heilman* et al. 1980). It is interesting to note that, according to this parameter, HPV 1 is more closely related to BPV 1 than to HPV 2 or 4 (*Heilman* et al. 1980); this may reflect surprising aspects of papillomavirus evolution.

The evaluation of the DNA sequences of HPV 1, HPV 6, and BPV 1 confirmed the colinearity and revealed a strikingly similar genome organization (*Chen* et al. 1982; *Danos* et al. 1982; *Schwarz* et al., to be published). All major open reading frames for proteins are in very similar positions and of comparable size (Fig. 2). One of the highly conserved regions lies in reading frame L 1, which was shown to code for the structural protein in the case of BPV 1 (*Heilman* et al., personal communication; see Sect. 6.3.2). The second highly conserved region in reading frame E 1 (Fig. 2) is supposed to code for early functions (*Heilman* et al. 1982; see Sect. 6.3.1).

It could be predicted from these data that the structural proteins should share common antigenic determinants. Papillomaviruses of man, cattle, dog, and rabbit had been compared previously by immunodiffusion in agar (*Le Bouvier* et al. 1966) and no cross-reactions between viruses from different species were shown. This finding was confirmed later by other techniques with native virus particles as antigen (*Lancaster* and *Olson* 1978; *Orth* et al. 1978b; *Pfister* et al. 1979b; *Pfister* and *Meszaros* 1980). The existence of group-specific antigens was finally disclosed by two lines of evidence:

1. Sera of rabbits, which bear a transplantable CRPV-induced carcinoma, react with the main structural polypeptide of disrupted HPV 1 virions and, in rare cases, with native HPV 1 virions (*Orth* et al. 1978a). The reactivity with disrupted virions may be explained by the fact that the carcinoma cells do not regularly produce virus particles but only unassembled structural proteins exposing additional antigenic determinants, which are masked in intact particles. Similarly, anti-HPV 1 polypeptide antisera react with CRPV antigens in immunofluorescence tests (*Orth* et al. 1978a). The reactivity of sera from carcinoma carriers with native virions indicates that further cross-reacting antigens on the virion surface provide a weak stimulus generally overlooked by the immune system and only realized by carcinoma carriers as a result of continuous antigenic stimulation.

2. On the basis of these observations antisera were raised against SDS-disrupted HPV 1 or BPV 1 particles and were tested with thin sections of warts from different animals, both by indirect immunofluorescence and by the peroxidase-antiperoxidase technique (*Jenson* et al. 1980; *Pfister* and *Meszaros* 1980; *Lancaster* and *Jenson* 1981). These sera reacted with lesions induced by HPV 1, 2, 3, and 5, BPV 1 and 2, and COPV. Virus-specific antigens were also detected in juvenile laryngeal papillomas (*Lack* et al. 1980; *Costa* et al. 1981; *Lancaster* and *Jenson* 1981) and condylomas (*Woodruff* et al. 1980; *Kurman* et al. 1981; *Morin* et al. 1981), a number of which were probably induced by HPV 6 or HPV 11. The antigen location corresponded to that of capsid antigens. The presence of virions was confirmed in some experiments by electron microscopy of antigen-positive areas (*Morin* et al. 1981).

The sera against SDS-disrupted viruses were negative with cells infected by polyoma-type viruses SV40, BK or polyoma, suggesting that papillomaviruses have a group-specific antigen unrelated to that of the second papovavirus subgroup. This is in line with the fact that even under relaxed conditions no cross-hybridization occurs between papillomavirus and polyoma virus DNAs.

Both cross-hybridization of DNAs under relaxed conditions and group-specific antisera provide useful tools to screen tumor tissues for papillomavirus fingerprints, even if the virus in question has not yet been character-ized and specific probes are not available (*Lancaster* and *Jenson* 1981).

4 Biology of Papillomavirus-Infection

Papillomaviruses cause epithelial or fibroepithelial proliferations of the skin or mucosa (Table 3). With most virus types the host range is limited to epithelial cells. Only few papillomaviruses are able to transform fibro-blast and affect the dermis. BPV 1 and 2 are the best characterized ones among them. Papillomavirus-induced tumors are primarily benign, show limited growth, and often regress spontaneously (*Massing* and *Epstein* 1963; *Rulison* 1942). Malignant conversion occurs with some types of lesions, mainly on the basis of long persistence.

4.1 Characteristics of Benign Tumors

Experimental infection with epitheliotropic viruses results in hyperplasia of cells in the spinous layer (acanthosis). The incubation period varies be-tween 3 and 18 months in the human system (*Rowson* and *Mahy* 1967); it is shorter in some animal systems. After experimental infection of rabbits with CRPV the incubation period lasts 3−8 weeks until papillomas appear (*Ito* 1975). There are no clear-cut ideas about the events taking place during these rather long incubation periods. It is generally assumed that virus infection leads to transformation of one or more basal cells, and trans-formation should result in increased proliferation and wart formation. Studies which were aimed at the localization of viral DNA replication in

Table 3. Localization and histopathology of papillomavirus-induced tumors

Tissue	Histopathology	Examples
Cutaneous Stratified epithelium	Acanthoma	HPV 1−5 HPV 7−10 HPV 12 HPV 14−15 BPV 3, 5
	Fibropapilloma	BPV 1, 2
	Fibroma	deer fibroma virus
Mucosa		HPV 6, 11, 13
	Acanthoma	BPV 4
	Fibropapilloma	BPV 1, 2

Data on animal viruses from *Shope* et al. (1958); *Lancaster* and *Olson* (1978); *Pfister* et al. (1979c); *Campo* et al. (1980, 1981). For references on HPV see Tables 4 and 5

CRPV-induced papillomas demonstrated virus-specific DNA in cell nuclei of the stratum granulosum and not in the stratum spinosum or germinativum (*Orth* et al. 1971). Also with HPV 1-induced human warts no labeling could be detected in the basal cell layer after in situ hybridization. Labeling started in the first or the second suprabasal cell layer (*Grussendorf* and *Zur Hausen* 1979), suggesting that extensive viral DNA replication first occurs in the stratum spinosum. The negative results with basal cells are easily explained by the relatively low sensitivity of the in situ hybridization technique, but nevertheless there is no direct evidence for the presence of papillomavirus DNA in cells of the stratum germinativum.

Analysis of the glucose-6-phosphate dehydrogenase phenotype in common warts of women who were heterozygous for this gene suggested a monoclonal origin of warts (*Murray* et al. 1971). *Friedmann* and *Fialkow* (1976), however, found both isoenzymes in each of four condylomata acuminata tested, thus indicating a multicellular origin.

Mature virus particles are first seen in association with the nucleoli of cells in the stratum spinosum (*Almeida* et al. 1962). In the stratum granulosum virions are spread throughout the nuclei and appear in paracrystalline arrays. After dissolution of cell structures, aggregates of virus are embedded in keratin in the stratum corneum. Virus-specific cytopathogenic effects are most prominent in the stratum granulosum.

In other words, epidermal cells are nonpermissive for papillomaviruses in the beginning of their differentiation process and become more and more permissive with increasing differentiation. This can theoretically be explained in two ways: Differentiated cells either produce a new product essential for virus replication, or lose a control mechanism which prevents virus multiplication. The molecular basis of this change is totally unknown. The events after papillomavirus infection of the epithelium are schematically summarized in Fig. 3.

The development of fibropapillomas was studied after experimental infection of cattle with cell-free virus from fibropapillomas (for review, see *Olson* et al. 1969). Within the first week after infection a fibroblastic reaction can be observed, which leads to massive fibroplasia during the next few weeks. Acanthosis and proliferation of the epithelium overlying the area of fibroplasia do not become visible before 4–6 weeks. Papillomavirus particles are demonstrable in the stratum granulosum and the stratum corneum; whereas it is interesting to note that no virus particles can be detected in the transformed proliferating fibroblasts. This holds true even for deer fibroma virus-induced neoplasms, which consist almost exclusively of a fibroma with a minimally hyperplastic epithelium.

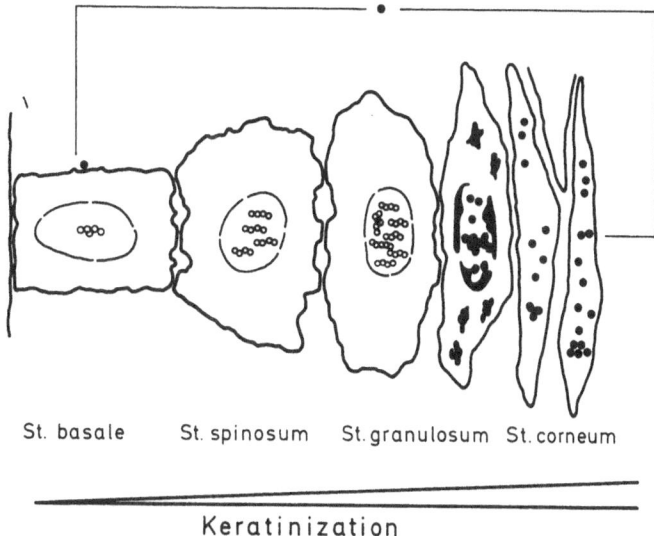

St. basale   St. spinosum   St. granulosum St. corneum

Keratinization

Cell proliferation
Viral replication
Maturation
Mature particles

Fig. 3. Schematic representation of the replication of papillomaviruses in epidermal cells. (○○○), viral DNA; (●●●), mature virus particle

4.2 Human Papillomavirus-Induced Tumors

All human papillomaviruses described so far induce pure epithelial proliferations. The tumors discussed below are regarded as HPV-induced, as a result of one or more of the following reasons:

1. Electron microscopic demonstration of virions in thin sections of tumor cells or in buffer extracts of ground tumor material.
2. Demonstration of virus-specific DNA by DNA–DNA or DNA–RNA hybridization.
3. Demonstration of papillomavirus group-specific antigens.
4. Experimental transmissibility of warts with cell-free viral extracts (see Table 4).

4.2.1 Clinical Aspects

On the skin there are the well-known plantar, common, and flat warts, showing acanthosis, prominent keratohyalin granules in the stratum granulosum, and hyperkeratosis as common features.

Epidermodysplasia verruciformis (Ev) was originally described as a genetic disease characterized by disseminated, persistent skin warts, usually arising during childhood, and by a high risk for developing skin cancer later in life (*Lewandowsky* and *Lutz* 1922). As will be discussed below, the patients show congenital defects of cell-mediated immunity, which make them prone to infections with certain types of HPV. The tumors resemble either flat warts, as frequently observed in children, or extremely flat warts, reddish plaques, and pityriasis versicolor-like lesions. The latter have a very typical histology, which led to the name of the disase: dysplastic swollen cells with pale staining cytoplasm, clustered in the stratum spinosum and granulosum (*Lewandowsky* and *Lutz* 1922; *Jablonska* et al. 1979; *Orth* et al. 1979).

On the mucosa, genital warts (condylomata acuminata) and laryngeal papillomas are both characterized by extensive acanthosis and papillomatosis, and little or no keratinization. A group of dysplasias of the cervix recently joined the family of HPV-induced mucosal tumors when they were shown to harbor a considerable number of virus particles in a few nuclei of the uppermost epidermal layers (*Laverty* et al. 1978; *Della Torre* et al. 1978; *Hills* and *Laverty* 1979; *Morin* and *Meisels* 1980). The presence of papillomaviruses was confirmed by demonstration of the group-specific antigen (*Shah* et al. 1980; *Woodruff* et al. 1980; *Kurman* et al. 1981; *Morin* et al. 1981) and by DNA hybridization (*Gissmann* et al. 1982a, 1983; *Kurman* et al. 1982). Dysplasia or koilocytotic atypia is a cytological or histological diagnosis on lesions which cannot be differentiated from cervical intraepithelial neoplasia with the colposcope. In view of their association with HPV, they are now referred to as flat condylomas or condylomata plana (*Meisels* et al. 1982).

Laryngeal papillomas appear in children and in adults. Juvenile lesions are often multiple, usually benign, but nevertheless a severe clinical problem if exuberant growth leads to obstruction of the respiratory tract. Papillomas of adults are usually solitary and represent precancerous lesions. They were earlier differentiated from juvenile laryngeal papillomas by their clinical features, by their different epidemiology (see Sect. 4.2.3), and by the lack of papillomavirus particles. However, by using the Southern blot hybridization technique, papillomavirus-specific sequences were recently disclosed, and the papillomavirus group-specific antigen was detected in two of the adult-onset papillomas (*Mounts* et al. 1982). Therefore, a differentiation on an etiological basis must be questioned.

Focal epithelial hyperplasia Heck was first described in Indians of the American continent (*Archard* et al. 1965). The disease also appears in other races but seems to be very rare in Caucasians (*Orfanos* et al. 1974; *Van Wyk* 1977). It occurs mainly in children and young adults, with frequent manifestation within the same family. Multiple slightly elevated

Table 4. Clinical characteristics of human wart diseases[a]

Type of papilloma	Demonstration of viral particles	Trans-missibility	Localization[b]	Histology	Malignant conversion
Verruca vulgaris	+ to +++ [c]	+	Back of hands and wrists	Hyperkeratosis, acanthosis, papillomatosis	2 case reports
Verruca plantaris	+++	+	Soles of the feet (and hands)	Hyperkeratosis, acanthosis, papillomatosis, inclusion bodies	–
Verruca plana	++	+	Face, hands	Hyperkeratosis, acanthosis	–
Epidermodysplasia verruciformis	++	+	Generalized	(See verruca plana) or hyperkeratosis, hypergranulosis, moderate acanthosis, large, clear cells	++++
Condyloma acuminatum	+	+	Genital and anal mucosa	Acanthosis, papillomatosis	+
Cervical flat wart	++	NT	Cervix	Acanthosis, koilocytosis	++
Laryngeal papilloma:					
juvenile	+	+	Larynx,	Acanthosis	+
adult	+		trachea	papillomatosis	+
Focal epithelial hyperplasia Heck	+	NT	Oral mucosa	Acanthosis	–

a For references see review articles by *Rowson* and *Mahy* (1967), *Steigleder* (1978), *Zur Hausen* (1977a). More recent references are given in the text.
b Only usual localizations are listed. Exceptions exist for nearly every tpye of papilloma.
c Number of (+) refers to quantity.
NT, not tested

papules appear on the red surface of the lips and on the labial and buccal mucosa. They may persist for several years but will not become malignant and finally tend to spontaneous remission. Papillomavirus particles were repeatedly detected in Morbus Heck lesions.

The characteristics of the various human wart diseases are summarized in Table 4.

Papillomavirus particles were further demonstrated, once in each lesion, in oral papillomas (*Frithiof* and *Wersäll* 1967), in a fibroma of the tongue (*Gross* et al. 1980), in an area of solar keratosis (*Spradbrow* et al. 1983), and in so-called multicentric pigmented Bowen's disease (*Kimura*

et al. 1978). Bowenoid papulosis is histologically similar to genital carcinoma in situ but biologically distinct, with another age distribution of the
patients and a multicentric origin of the lesions. HPV 6 DNA was detected
once, and hybridization under relaxed conditions revealed HPV-specific
DNA in another case, where further classification was not possible (*Zachow*
et al. 1982). The exact role of papillomaviruses in these lesions remains to
be established.

4.2.2 HPV Type-Specific Effects

Distinct clinical symptoms are each clearly associated with a limited number of individual HPV types (Table 5). It is noteworthy that viruses which
were shown to be related by DNA cross-hybridization (see Table 1) again
form groups according to their biological properties. For example, HPV 6,
11, and 13 (see Table 2) all affect the mucosa, but usually at different
sites and leading to lesions with different morphology and histology. The
group of HPV 5, 8, 9, 12, 14, and 15 is even more homogeneous, infecting
only patients with Ev and leading to very characteristic pityriasis versicolor-like lesions. However, wart morphology alone would certainly be
inadequate to determine the type of HPV. For example, common warts
may be induced by totally different viruses like HPV 1, 2, 4, and 7.

A more specific picture emerges from wart histology (*Grussendorf*
1980; *Jablonska* et al. 1980; *Orth* et al. 1981; *Gross* et al. 1982):

The features of HPV 1-induced warts correlate very well with those
described for myrmecia warts: extensive papillomatosis, eosinophilic cytoplasmic keratohyalin inclusions, which become confluent and very large in
the upper epidermal layers, and basophilic nuclear inclusions, which were
earlier shown by electron microscopy to correspond to paracrystalline
aggregates of virus particles (*Almeida* et al. 1962). There is pronounced
hyperkeratosis and parakeratosis.

For a number of HPV types specific cytopathogenic effects have been
described: The most prominent feature of HPV 2-induced warts are the
numerous keratohyalin granules of varying size, shape, and stainability.
HPV 3-induced warts are characterized by moderate or almost no papillomatosis and by extensive vacuolization: Pyknotic nuclei are centrally
located in large perinuclear vacuoles surrounded by a ring of keratohyalin granules, giving the impression of bird's eyes (*Kaufmann* et al. 1978;
Laurent et al. 1978). Clusters of large clear cells are observed in the other-
wise inconspicuous stratum granulosum of HPV 4-induced warts. The
crescent-shaped nuclei are peripherally located. HPV 5-induced lesions are
very characteristic, with basophilic foamy giant keratinocytes first detectable in the lower malpighian layers. HPV 6-induced condylomata acuminata reveal marked perinuclear vacuolization with marginal sickle-shaped

Table 5. Association between human papillomavirus types and clinical symptoms

Type of papilloma	HPV type															
	1	2	3	4	5	6	7	8	9	10	11	12	13	14	15	16
Verruca vulgaris	+	+++		++			+									
Verruca plantaris	+++	+		+												
Verruca plana			++							++						
Ev flat wart			++		+					++						
Ev pityriasis-like lesion								+	+					+	+	
Condyloma acuminatum	+	+				+++					++	+				
Cervical flat wart						+					++					+
Laryngeal papilloma											++					+
Focal epithelial hyperplasia Heck	+												++			

The number of (+) reflects the prevalence of different virus types in the various papillomas. The data are summarized from *Orth* et al. (1977, 1978b, 1981, personal communication), *Pfister* and *Zur Hausen* (1978a, b), *Krzyzek* et al. (1980); *Petzoldt* and *Pfister* (1980), *Ostrow* et al. (1981), *Gissmann* et al. (1982a, 1983), *Dürst* et al. (1983), *Pfister* et al. (1983b) nuclei in the malpighian and granular layer. HPV 7-induced warts show large clear cells with centered nuclei and heavily staining cells with kerato-hyalin granules in close proximity.

The amount of virus particles in various warts differs considerably, ranging from less than 10^5 particles in laryngeal papillomas (*Boyle* et al. 1973) to 10^{12} particles in some plantar and common warts (*Barrera-Oro* et al. 1962). The efficiency of virion production seems to be a property of the HPV type (see Tables 4 and 5). The most abundant particle synthesis by HPV 1 is the reason why this type was taken as the only representative of HPV for a long time. Virus production may be influenced, however, by the genetic constitution of the host, as seen with the Shope papillomavirus: It induces papillomas with high particle yields in cottontail rabbits *(Sylvilagus floridanus)*, but there are hardly any virions detectable in papillomas of the domestic rabbit *(Oryctolagus cuniculus)* induced by the same virus stock (*Friedewald* and *Kidd* 1944; *Ito* 1971).

4.2.3 Epidemiology of HPV Infections

4.2.3.1 Incidence of Warts
Plantar, common and flat warts mainly occur in children and young adults (*Zur Hausen* et al. 1975; *Pfister* and *Zur Hausen* 1978b). The low incidence later in life is probably due to immune mechanisms (see Sect. 7). Condylomas show a completely different age distribution which is easily explained by the venereal mode of transmission (*Oriel* 1971a, b). They are prevalent in populations of high sexual promiscuity and their peak incidence coincides with the maximum of sexual activity (*Zur Hausen* 1977a). Laryngeal papillomas have a bimodal age distribution, with a first peak between 2 and 5 years and a second one between 40 and 60 years. They are referred to as juvenile and adult lesions. From clinical observations it was suspected that infection of children may occur during delivery from a mother with condylomas. In this connection it is noteworthy that flat condylomas of the cervix and laryngeal papillomas were found to harbor the same type of HPV, namely HPV 11 (*Gissmann* et al. 1982b, 1983). Adult laryngeal papillomas are associated with closely related viruses (*Mounts* et al. 1982). It is not yet clear if they are induced by new infections or by reactiviation of a virus, which might persist after clinically apparent or inapparent infection during childhood (see Sect. 4.3).

4.2.3.2 Efficiency of Papilloma Induction
As deduced from experimental systems the efficiency of papilloma induction is rather low (*Bryan* and *Beard* 1940). Furthermore, it certainly varies with individual virus types. One study demonstrated that approximately 60% of sexual partners of infected individuals developed genital

warts (*Oriel* 1971a). On the other hand, family members of Ev patients did not develop warts, in spite of intimate contact (*Jablonska* et al. 1980).

HPV 5, 8, 9, 12, 13, 14, and 15 were only detected in very special patient groups, which points to an important role of host response. HPV 13, which is associated with focal epithelial hyperplasia Heck, may be limited to certain ethnic groups, and the reason for this is unknown (*Pfister* et al. 1983b). The other examples are confined to Ev patients and seem to depend on a reduced immune reactivity, for at least a HPV 5 infection was also observed in an immunosuppressed renal transplant recipient (*Lutzner* et al. 1980). In the case of other types, like HPV 2 or 3, the efficiency of wart induction is at least significantly increased by immunosuppression of the host (*Jablonska* et al. 1980). From this evidence alone the sudden onset of multiple HPV 2-induced warts at higher age can be taken as a possible indication for an impairment of the immune system, as the result, for example, of chronic lymphatic leukemia or Hodgkin's disease (*Gross* et al. 1983).

4.2.3.3 Humans and Animal Papillomaviruses

Papillomaviruses show a remarkable host specificity, and all attempts to transfer HPV to other species have failed (*Rowson* and *Mahy* 1967). In contrast, BPV 1 and 2 show a rather wide host range when compared with other papillomaviruses (see Sects. 4.4.4 and 5.2). This has led to repeated speculations on transmission to man, in spite of the different histology of human warts and in spite of the negative results of in vitro transformation experiments with human cells (*Black* et al. 1963). Support came from epidemiological surveys that revealed the prevalence of virus-induced warts among butchers and veterinary surgeons (*Bosse* and *Christophers* 1964; *De Peuter* et al. 1977). However, seroepidemiological studies using a solid phase radioimmunoassay failed to demonstrate antibodies against BPV 1 and 2 in man, in contrast to cattle, where 19% of a nonselected group were anti-BPV 1 positive (*Pfister* et al. 1979b). Direct virological analysis of 64 warts from butchers (*Orth* et al. 1981; *Ostrow* et al. 1981) was also negative for BPV 1 or BPV 2. HPV 7 was isolated during these studies. It showed no relationship with known human or bovine papillomaviruses and was very prevalent in this patient group (*Orth* et al. 1981). As this virus type could be isolated only from people dealing with meat (*Orth*, personal communication), the possibility cannot be excluded that it represents a so-far unknown animal papillomavirus.

4.3 Inapparent Infections and Virus Persistence

At the moment little is known about clinically inapparent primary infections and about the persistence of a virus without clinical symptoms. It has been well established by seroepidemiology that people without any known wart history may have papillomavirus-specific antibodies (*Cubie* 1972; *Pfister,* unpublished work) and cell-mediated immune reactions (*Lee* and *Eisinger* 1976); however, inconspicuous lesions may escape attention, so that these observations are not proof of inapparent infections.

There are some arguments in favor of symptomless virus persistence:

1. Condylmoas were often noted during pregnancy (*Cook* et al. 1973) or following severe liver disease, without venereal contact with infected persons within the normal incubation period.
2. Papillomavirus particles were detected in the midstream urine of pregnant women who had no evidence of any genital or other warts (*Lecatsas* and *Boes* 1979). Furthermore, genital warts have been reported in newborns whose mothers show no clinical evidence of disease (*Felman* and *Nikitas* 1979).
3. By means of the Southern blot technique, papillomavirus DNA was detected in clinically inconspicuous mucosa of the larynx from patients who had no laryngeal papilloma recurrences for 2 years (*Steinberg, Abramson,* and *Topp,* personal communication).
4. Common warts disappeared after treatment with Tigason. Histologically there was no longer any evidence for virus-specific cytopathogenic effects, and no viral DNA was detected by the Southern blot technique. When treatment was stopped, the warts recurred and harbored the same HPV 2 subtype as before, as deduced from the restriction enzyme cleavage patterns of its DNA (*Gross* et al. 1983). In view of the great number of HPV 2 subtypes, reactivation of a persisting virus seems more likely than a new, independent infection.

In humans nothing is known so far about the cells in which the virus could persist or about the mode of viral persistence. The molecular basis might be quite similar to that of virus persistence during the incubation period. Indeed it may be merely semantic in some cases to differentiate between the activation of a persisting virus and the extremely long latency period after infection. An understanding of one problem will certainly help to elucidate the other.

Domestic rabbits provide a beautiful animal model for papillomavirus persistence in apparently healthy tissue (*Parsons* and *Kidd* 1942). They suffer from oral papillomas on the undersurface of the tongue, which are induced by a virus unrelated to CRPV. The virus is readily recoverable from mouth washings of tumor-bearing animals and can be demonstrated by its biological activity in inoculation experiments. Small amounts of virus were detectable in washings from animals which were completely healthy at the time but developed papillomas later in life. This is suggestive of persistent infection with occasional spontaneous virus production. Mouth washings of animals which neither had a papilloma history nor developed tumors later on were negative. Application of tar was shown to induce tumor formation in domestic rabbits.

It is interesting to note that tar treatment did not elicit oral papillomas in cottontail rabbits, which are principally highly susceptible to the virus. This may indicate that there is no persistent infection in the wild animals. So far, this system has not been examined at the molecular level.

The multimammate mouse *(Mastomys natalensis)* represents another fascinating model system for persistence of latent papillomavirus genomes *(Amtmann,* personal communication). Episomal MnPV genomes were detected by Southern blot hybridization in histologically normal adult liver, colon, muscle, and skin tissues, and even in embryos. The developmental stage at which congenital infection occurs is not clear at the moment. In skin tissues the viral DNA content increased from two copies per cell at 16 weeks of age to several thousand copies at 64 weeks of age. The first tumors were noted at the age of 1 year. Expression of the viral genomes was detected in tumors. In the normal skin, gene expression, if any, was below test sensitivity. Further analysis of this system may provide valuable information on the persistence mechanisms of papillomaviruses.

4.4 Malignant Conversion

A number of papillomaviruses induce tumors that may eventually progress to carcinomas. This was first noted and studied in depth with CRPV *(Rous* and *Beard* 1935; *Kidd* and *Rous* 1940; *Syverton* 1952): 25% of infected cottontail rabbits suffer from invading squamous cell carcinomas, which arise on the basis of papillomas within several months. They metastasize to the lungs and other organs, eventually causing the death of the animals.

Further examples from the animal kingdom are BPV 4-induced esophageal papillomas of cattle *(Jarrett* et al. 1978), virus-induced bovine ocular papillomas *(Spradbrow* and *Hoffmann* 1980; *Ford* et al. 1982), and papillomas in sheep *(Vanselow* and *Spradbrow* 1982). They are all regarded as precursor lesions to squamous cell carcinomas. Finally, MnPV-induced papillomas may progress to keratoacanthomas and sqamous cell carcinomas *(Reinacher* et al. 1978).

Conversion of human papillomas into squamous cell carcinomas has been reviewed by *Zur Hausen* (1977a). It holds true for Ev lesions, condylomata acuminata, condylomata plana, and laryngeal papillomas (see Table 4).

Experimental analysis of the CRPV system revealed a number of general features, which will be covered first, as they seem to be valid for most other systems also.

4.4.1 Papilloma-to-Carcinoma Sequence in Rabbits

Carcinoma induction is not an early consequence of CRPV infection but usually occurs during long persistence of papillomas. The critical period is about the twelfth month after experimental infection of cottontail rabbits (*Syverton* 1952). This implies that malignant conversion is either due to an unlikely event or to a multistep process. Finally, however, carcinoma development seems almost inevitable: Papillomas rarely continue as benign growths for more than 18 months.

Small pieces of newly induced papillomas were experimentally transferred to inner organs of rabbits (*Rous* and *Beard* 1934). They proliferated actively, were markedly invasive and destructive, and led to the death of the animals. This indicates that some characteristics of malignancy appear early in papilloma development.

Experimental CRPV infection of domestic rabbits leads to papillomas, which persist in more than 90% of the animals. This contrasts with the natural infection of cottontail rabbits, where about 50% of the warts regress within 6–12 months, but is in line with observations made after experimental infection of the natural host (*Syverton* 1952). Approximately 75% of domestic rabbit papillomas become malignant after about 9 months, which corresponds to a threefold higher incidence when compared with cottontail rabbits, where 25% of wart carriers develop carcinomas, irrespective of natural or experimental infection. This difference in malignant transition rates points to the importance of host reactivity. In the natural host the system is significantly better balanced than in the "artificial host."

The synergistic effect of extrinsic cofactors was investigated by *Rous* and his co-workers. Methylcholanthrene or tar was repeatedly supplied to domestic rabbit papillomas and they became malignant earlier and at more sites (*Rous* and *Friedewald* 1944). Concurrent exposure to hydrocarbons and virus had the same effect (*Rogers* and *Rous* 1951). Most interestingly, the system also works in reverse: Repeated tarring of the skin led to benign tumors, which rarely evolved into carcinomas; subsequent intravenous inoculation of the virus resulted in high yields of carcinomas and papillomas (*Rous* and *Kidd* 1938). On the basis of this evidence there can be little doubt that CRPV plays a synergistic role in carcinogenesis.

The viral DNA continues to be associated with the carcinomas (*Stevens* and *Wettstein* 1979; *Wettstein* and *Stevens* 1980). Two transplantable carcinoma lines (Vx 2 and Vx 7) were kept in animals for 30 years and they still harbor viral DNA (*Favre* et al. 1982; *McVay* et al. 1982). It would be interesting to try to cure the carcinomas from viral DNA (see Sect. 9.2) and see if the cells are still malignant. This would provide evidence that the persistence of viral DNA is essential for the maintenance of the malignant state – or not. The tumor-inducing potential of the persisting DNA seems to be unchanged. Transfection of the skin of domestic rabbits with DNA extracts from carcinomas primarily leads to papillomas and not directly to carcinomas (*Ito* 1963). This is the same pathway as in transfection experiments with DNA from virus particles.

Infectious virus has never been recovered from primary carcinomas or their metastases (*Kidd* and *Rous* 1940). However, animals bearing transplanted carcinomas reveal high titers of antibodies against virus capsid antigen. Minute amounts of antigen were actually detected in such tumors, and some infectious virus could also be isolated (*Rogers* et al. 1960).

4.4.2 Bovine Alimentary Tract Carcinomas

In Scotland there is a high-incidence area of bovine alimentary tract carcinomas, which occur in close proximity to BPV 4-induced papillomas (*Jarrett* et al. 1978; *Campo* et al. 1980). In some cases progression from a papilloma to a carcinoma was histologically demonstrated within the same lesion (*Jarrett* et al. 1978). In this system, virus infection alone does not lead to carcinomas; they were observed only in cattle in combination with a bracken fern diet. This plant contains a radiomimetic substance not identified so far, and long-term diet is thought to be immunosuppressive and carcinogenic. It has not been possible up to now to demonstrate BPV 4 DNA in the carcinomas (*Jarrett,* personal communication), which is in contrast to the CRPV system. Further studies must show if this reflects a different role of BPV 4 in carcinogenesis, or a different mechanism, or both.

4.4.3 Human Malignancies

4.4.3.1 Epidermodysplasia verruciformis
About one-third of Ev patients develop cancer between 2 and 60 years after the onset of verrucosis, on average after 24 years (*Lutzner* 1978). The carcinomas are often of the in situ type but also grow invasively with heavy destruction of the surrounding tissue, eventually leading to the death of the patient. They hardly ever metastasize.

The carcinomas are found mainly at light-exposed sites such as the forehead (*Jablonska* et al. 1972). This seems to provide another example of synergistic effects between papillomavirus infection and extrinsic factors – here most probably UV light. It is interesting that Ev has a relatively good prognosis in Africans, as compared with Caucasians (*Jacyk* and *Subbuswamy* 1979), which may be due to the protective role of skin pigmentation. Africans are infected by the same virus types as Caucasians (*Pfister* et al. 1981b; *Blanchet-Bardon, Lutzner, Puissant,* and *Orth,* personal communication). In rare cases Africans develop carcinomas, and there is one report of a carcinoma at the scrotum, which cannot be regarded as sun exposed (*Blanchet-Bardon* et al., personal communication). This may point to other cofactors or to "spontaneous" malignant conversion of Ev papillomas.

An etiological role of papillomaviruses in the progression from warts to carcinomas was first suggested by the observation that malignant conversion preferentially occurred in connection with pityriasis-like warts induced by HPV 5 or 8 (*Orth* et al. 1979). Twelve patients suffering from HPV 5 and 8 infection all developed carcinomas, whereas no malignancies were observed in connection with HPV 3-induced flat warts (*Laurent* et al. 1978; *Kienzler* et al. 1979; *Orth* et al. 1979; *Pfister,* unpublished work).

The role of papillomaviruses in Ev carcinogenesis was substantiated when HPV 5 DNA was detected in the carcinomas themselves (*Orth* et al. 1980; *Ostrow* et al. 1982; *Pfister* et al. 1983a). The three HPV 5 DNAs were characterized in more detail by restriction enzyme analysis and proved to be very similar. To exclude the possibility that the viral DNA was due to contaminating papilloma material, one carcinoma was examined histologically and DNA was prepared from tissue sections following immediately (*Pfister* et al. 1983a). In one case DNA could be extracted from a rare metastasis, which was also positive (*Ostrow* et al. 1982). In situ hybridization studies revealed only a few cells (approximately 1 in 10 000) that contained abundant viral DNA sequences (*Orth* et al. 1980). The absence of silver grains from other nuclei does not exclude the presence of a few viral genome copies, which would not be detected by this method. The in situ hybridization result suggests an uneven distribution of viral DNA as a result of increased replication in some cells. Even there the viral cycle remains abortive, however. No virus particles were detectable with the electron microscope in the labeled nuclei (*Orth* et al. 1980). This is in line with earlier observations that carcinomas are free of particle synthesis, whereas virions are readily demonstrated in the benign lesions (*Aaronson* and *Lutzner* 1967; *Jablonska* et al. 1970, 1972; *Ruiter* and *van Mullem* 1970) – a situation that parallels the CRPV system! Further carcinomas were analyzed in the meantime (*Orth* et al., personal communication), and HPV 5 was detected in ten carcinomas of eight patients, HPV 8 in four carcinomas of two patients, and HPV 3 once. The limited number of HPV types in carcinomas contrasts with the broad heterogeneity of HPV in Ev papillomas (see Table 5), and it is tempting to speculate that only some HPV types are endowed with a cancerigenic potential. In view of the wide distribution of HPV 3 in flat warts, it is of great interest that this virus was also detected in a carcinoma although its association with malignant tumors is very weak both from epidemiological data and from the relative number of HPV 3-positive carcinomas.

Ev itself is a very rare disease, but basically similar conditions may be found in immunosuppressed patients (see Sects. 4.2.3 and 7.2.2); they are known to have an increased incidence of warts and skin cancer (*Spencer* and *Andersen* 1970; *Starzl* et al. 1970; *Koranda* et al. 1974; *Mullen* et al. 1976; *Hoxtell* et al. 1977). About 42% of the patients develop warts with a lag phase of more than 1 year after transplantation, and de novo squamous cell carcinomas are about 35 times as frequent as in the general population.

Flat warts induced by HPV 3 are most prevalent (*Pfister* et al. 1979a). Recently, HPV 5 infections have been reported in two renal allograft recipients (*Lutzner* et al. 1980), and in one of these patients multiple in situ cancers were noted. Anti-HPV 8 antibodies were detected in an immunosuppressed organ recipient and in one patient undergoing hemodialysis (*Pfister* 1980). It would be of great interest to evaluate the role of HPV 5, 8, and 3 in the pathogenesis of skin cancers in the transplant patient population.

4.4.3.2 Genital Cancer

The epidemiology of human genital cancer clearly suggests that infectious events play a role in the etiology of the disease (*Rotkin* 1973). A number of infectious agents have been incriminated, and papillomaviruses were added only recently (*Zur Hausen* 1976). They soon gained ground, however, when koilocytotic atypia lesions of the cervix were shown to harbor HPV (see Sect. 4.2.1). These so-called flat condylomas frequently coexist in close proximity to or even intermingled with dysplastic and neoplastic epithelium. HPV-containing lesions are associated with 25%–50% of dysplastic and neoplastic processes (*Syrjänen* 1979; *Meisels* et al. 1982). Follow-up studies of condylomas of the cervix revealed a progression to moderate dysplasia, carcinoma in situ, and invasive carcinoma in 10% of the patients with flat condylomas (*Meisels* et al. 1982).

In contrast to flat condylomas, condylomata acuminata mainly involve the vulva, penis, perineal, and perianal areas and show a low incidence of malignant conversion. The nevertheless considerable number of case reports has been reviewed by *Zur Hausen* (1977a). According to these reports at least 5% of carcinomas of the vulva arise within persisting genital warts. Usually there is a long latency period of more than 10 years between the appearance of condylomas and malignant conversion. The peak incidence of vulval and penile carcinomas is approximately 30 years after the peak incidence of condylomata acuminata. Epidemiological studies from Uganda have indicated a significant correlation between the frequency of condylomas in certain tribes and the incidence of carcinomas of the vulva and penis (*Schmauz* and *Owor* 1980).

Giant condylomata acuminata, originally described by *Buschke* and *Löwenstein* (1931), represent a special pathological entity with invasive growth properties. These verrucous carcinomas usually do not metastasize, but there are reports in the literature of progression into metastasizing squamous cell carcinomas (*Zur Hausen* 1977a).

HPV-specific DNA could recently be demonstrated in cervical carcinomas, vulval carcinomas, and Buschke-Löwenstein tumors (*Green* et al. 1982; *Zachow* et al. 1982; *Gissmann* et al. 1983; *Dürst* et al. 1983).

Two of 31 cervical carcinomas harbored sequences related to papillomavirus DNA, which had been cloned from Ev lesions (*Green* et al. 1982). It probably represents HPV 10 (*Green,* personal communication). This "Ev virus" was also disclosed in two condylomas, which indicates that it is not restricted to skin warts but infects the genital area also. HPV 11 DNA was detected in five out of 27 cervical cancers, in three invasive ones and in two carcinomas in situ (*Gissmann* et al. 1983). HPV 16 DNA was identified by hybridization under relaxed conditions and was directly cloned from a biopsy of an invasive cervical carcinoma (*Dürst* et al. 1983). It showed less than 0.1% cross-hybridization. Under stringent conditions HPV 16 DNA hybridized to a high percentage of cervical carcinoma DNAs (11/18 biopsies from German patients and 8/23 biopsies from Africa and South America). In contrast, HPV 16 was found only twice among 33 condylomata acuminata and twice among 20 cervical dysplasias.

Among the three HPV types detected in cervical carcinomas, HPV 10 is a newcomer to the genital area, and its prevalence remains to be established. The presence of HPV 11 fits nicely with its predominance in flat condylomas (*Gissmann* et al. 1983), which have already been described as being closely associated with carcinomas (see above). The absence of HPV 6 is not too surprising, as it prevails in condylomata acuminata (70%) and was more rarely detected in condylomata plana (*Gissmann* et al. 1983). The "preference" of HPV 16 for malignant tumors, compared to its occasional presence in benign lesions, deserves special interest; it provides a strong argument against contamination of the carcinoma material with HPV DNA from adjacent papillomas (*Dürst* et al. 1983).

HPV 10 and HPV 16 were also detected in two vulval carcinomas each, and HPV 16 DNA hybridized to one penile cancer DNA (*Green* et al. 1982; *Dürst* et al. 1983). A HPV 6 probe hybridized to one carcinoma in situ of the vulva under stringent conditions, but that experiment allowed no differentiation between HPV 6 and HPV 11 (*Zachow* et al. 1982).

Six out of seven Buschke-Löwenstein tumors were shown to harbor DNA of HPV 6 or HPV 11, i.e., of those types which are also prevalent in condylomata acuminata and plana (*Gissmann* et al. 1982a; *Boshart,* personal communication). DNAs from two verrucous carcinomas of the vulva also hybridized to HPV 6 DNA under stringent conditions (*Zachow* et al. 1982).

Certainly the demonstration of HPV DNA in genital cancer biopsies does not yet prove a viral etiology of these tumors. However, as observed with Ev carcinomas, there is also a preferential association between genital carcinomas and certain HPV types, especially HPV 16. This may point to an increased cancerogenic potential and to an etiological role of this type. It would be desirable to reveal the viral DNA in the individual cancer cells, in a metastasis, or in a carcinoma-derived cell line. Furthermore, the latter would provide interesting experimental possibilities.

4.4.3.3 Laryngeal Carcinomas

Malignant conversion of juvenile laryngeal papillomas seems to be an extremely rare event (for reviews see *Zur Hausen* 1977a). In addition to laryngeal carcinomas there is a report of a bronchogenic squamous carcinoma in a 19-year-old man with recurrent laryngeal papillomatosis since the age of 4 (*Siegel* et al. 1979). X-irradiation of recurrent laryngeal or tracheal papillomas of children or young adults led to a considerable number of carcinomas after 5–40 years (reviewed by *Zur Hausen* 1977a), indicating a synergistic effect between X-rays and papillomavirus infection.

Adult laryngeal papillomas represent a clearly premalignant lesion with rates of malignant transition above 20% (*Kleinsasser* and *Oliveira e Cruz* 1973). Heavy smoking probably acts as the promoter of carcinoma formation. Preliminary attempts to detect HPV 11 DNA (isolated from laryngeal papillomas) in carcinomas of the larynx were unsuccessful (*Gissmann* et al. 1982b).

4.4.3.4 Plantar and Common Warts and Malignancies

A more accurate heading would be: "Plantar and Common Warts and No Malignancies." There are two case reports on carcinomas arising from the basis of a common wart (*Grussendorf* and *Gahlen* 1975; *Shelley* and *Wood* 1981). Unfortunately, the virus was not typed in these studies. In view of the ubiquity of common warts, this number is exceedingly small. Plantar and common warts may persist for many years; they are obviously exposed to mechanical factors, trauma, and sunlight, which are all suspected of being cofactors of carcinogenesis. Therefore, the situation with plantar and common warts suggests that malignant conversion depends considerably on the infecting virus type. DNAs of HPV 1 and HPV 2, which are most prevalent in these lesions, were used to screen 156 and 145 human cancers, respectively, for papillomavirus sequences, with negative results (*Green* et al. 1981). This evidence confirms that both viruses are unlikely to be associated with human cancer, and that the induced lesions are really benign.

4.4.4 Equine Sarcoids

Sarcoids are the most common spontaneous tumors of horses. They are not likely to metastasize but are locally invasive and usually recur after surgical removal. BPV 1- or 2-specific DNA was repeatedly demonstrated in naturally occurring equine sarcoids (*Lancaster* et al. 1979; *Amtmann* et al. 1980; *Pfister* and *Kaaden*, unpublished work). The etiological role of these viruses can be substantiated by earlier transmission experiments: Experimental infection of horses with bovine fibropapillomavirus led to typical sarcoids (*Olson* and *Cook* 1951). Papillomaviruses thus offer an impressive example for the naturally occurring induction of a nonproductive, semimalignant tumor in an alien, nonpermissive host – a phenomenon which is well known in experimental tumor virology, where viruses such as SV40 or polyoma are only tumorigenic in the alien, nonpermissive host (*Tooze* 1980).

From cultured cells of an equine sarcoid, a C-type retrovirus was recently detected which elicits rapid morphological transformation of primary equine dermal fibroblasts (*Fatemi-Nainie* et al. 1982). It will be interesting to clarify whether this reflects an alternative pathway to equine sarcoids or a possible interaction between papillomaviruses and retroviruses in the genesis of equine sarcoids.

5 Cell Culture and Animal Model System

In spite of many efforts to grow papillomaviruses in vitro, so far there exists no reproducible permissive culture system. Abortive infection could be demonstrated in some cases, but a clear-cut transforming ability was only shown for BPV 1 and 2.

5.1 Human Papillomaviruses

Infection of fetal rabbit kidney cells or fetal human lung fibroblasts with HPV from plantar warts leads to a transient stimulation of DNA synthesis (*Butel* 1972; *Lancaster* and *Meinke* 1975). No cytopathic effects were observed with either cell system, but in human fibroblasts viral DNA sequences were detectable at low levels for at least 30 doublings after infection (*Lancaster* and *Meinke* 1975).

Rheinwald and *Green* (1975) developed a method for long-term cultivation of human epidermal keratinocytes. The cells partially differentiate in vitro leading to cornified cells which do not synthesize DNA any longer. Such cultures, which closely resemble the natural target tissue of HPV, could be infected with virus from plantar warts, and the viral DNA persisted and replicated as a stable episome (*LaPorta* and *Taichman* 1982). The high genome copy number of 50–200 per cell indicates that viral DNA replication in keratinocytes is much more efficient than in fibroblasts. No capsid protein production was detectable, however, and attempts to isolate virus particles from infected cultures were unsuccessful. Nevertheless, this system is likely to provide a useful model for studying early events of HPV infection, but the degree of in vitro differentiation is obviously insufficient for productive infection. In order to circumvent this problem, human skin grafts to nude ar antithymocyte serum-treated mice were infected by HPV in earlier experiments (*Pass* et al. 1973; *Cubie* 1976). No warts developed within the observation period, but this was not too amazing in view of the sometimes very long incubation period in humans.

5.2 Transformation by BPV 1 and 2

Two experimentally important properties of BPV 1 and 2 depend on their ability to stimulate fibroblasts:

1. Transmission to alien hosts, especially to laboratory animals, always involves infection of the fibroblasts.
2. All cells transformed in vitro and used for biochemical studies are fibroblasts.

Much of our present knowledge about the molecular biology of papillomavirus infection (see Chap. 6) is derived from these two model systems. Therefore, the biological properties will be outlined in more detail, and it should be borne in mind that BPV-transformed fibroblasts do not represent the usual virus-cell system for papillomaviruses.

As far as laboratory animals are concerned, connective tissue tumors could be induced in C_3H/eB mice (*Boiron* et al. 1964) and in the lagomorph *Ochotona rufescens* (*Puget* et al. 1975). Syrian hamsters develop fibromas and fibrosarcomas, meningiomas or chondromas, depending on the route of inoculation (*Friedmann* et al. 1963; *Boiron* et al. 1964; *Cheville* 1966; *Robl* and *Olson* 1968).

Hamster and mouse tumors were examined by electron microscope and by infectivity assays and proved to be free of mature viruses (*Lancaster* et al. 1979). BPV 1 or 2 persist on the DNA level (*Moar* et al. 1981a; *Pfister* et al. 1981a). All the experimental tumors grow progressively. Those from ochotona and from hamsters are transplantable (*Robl* and *Olson* 1968; *Puget* et al. 1975), but only hamster tumors metastasize to internal organs in about 10% of the cases (*Robl* and *Olson* 1968).

In vitro, BPV 1 and 2 are able to transform mouse cell lines NIH 3T3 and C 127 to focus formation, growth in soft agar, and to tumorigenicity in the nude mouse (*Dvoretzky* et al. 1980). Transformation follows single-hit kinetics. It can also be achieved by transfection with isolated viral DNA (*Lowy* et al. 1980). Primary hamster embryo cells can be transformed to growth in soft agar both by virus infection and DNA transfection (*Morgan* and *Meinke* 1980). On the basis of these data it is most probable that BPV 1 and 2 were also responsible for the results of earlier experiments on the oncogenicity of bovine papillomaviruses, which were carried out with untyped virus material from bovine papillomas: The virus was shown to transform primary fetal bovine skin cells (*Thomas* et al. 1963, 1964; *Boiron* et al. 1964) and, even more efficiently, fetal cells from the conjunctiva, the palate, and the vascular meninges (*Meischke* 1979). Transformation of secondary fetal skin cells was achieved by transfection with phenol-extracted BPV DNA (*Boiron* et al. 1965). The criteria for transformation were altered cell morphology (cells become long, thin, spindle-shaped), and increase of growth rate and life span. By criteria such as morphological changes, acid formation, and increased growth rate, cell lines from fetal bovine conjunctivae and kidneys were also shown to be transformed by BPV (*Black* et al. 1963). As far as cells from heterologous species are concerned, primary embryo cells from C_3H/eB, C 57/BL and Balb/C mice (*Thomas* et al. 1964), could be transformed by the above-mentioned criteria. Hamster embryo cells infected with BPV are morphologically altered (*Black* et al. 1963); they grow in soft agar (*Morgan* and *Meinke* 1980) and are tumorigenic in hamsters (*Geraldes* 1969). Transformation assays with primary cultures of human embryonic skin-muscle or kidneys were negative, as were assays with several human fibroblastic strains (*Black* et al. 1963; *Dvoretzky* et al. 1980).

It should be noted that among BPV-infected cells, which are transformed by one parameter or the other, only hamster cells were shown to be tumorigenic in the animal (*Geraldes* 1969). The tumors which arise after subcutaneous injection of BPV 1-transformed embryo fibroblasts still harbor BPV DNA (*Amtmann* and *Pfister,* unpublished work), thus underlining the viral role in tumorigenesis.

Transformed bovine cells were oncogenic neither in calves (*Black* et al. 1963; *Meischke* 1979) nor in hamsters (*Boiron* et al. 1964), and transformed mouse cells produced no tumors in isologous mouse strains during an observation period of 2 months (*Thomas* et al. 1964). Similarly surprising is the rather benign character of the cells which were established from a bovine fibroma and from equine fibrosarcomas. They are contact inhibited, do not grow in soft agar, and do not form tumors in the nude mouse (*Lancaster* 1981). In contrast, BPV 1-infected mouse cell lines NIH 3T3 and C 127 are transformed by all these parameters (*Dvoretzky* et al. 1980), and cell lines from hamster fibromas are highly oncogenic in the Syrian hamster (*Breitburd* et al. 1981). This all points to a rather complex virus—cell—animal interaction in some systems.

5.3 Transformation by BPV 4

Quite recently, *Campo* and *Spandidos* (1983) succeeded in transforming NIH 3T3 mouse fibroblasts by transfection with molecularly cloned BPV 4 DNA. The transformed cells lost contact inhibition, were anchorage independent, required low serum, and were tumorigenic in nude mice. These results deserve special interest as BPV 4 causes pure epithelial lesions in vivo and is believed to be associated with esophageal carcinomas of cattle.

6 Molecular Biology of Papillomavirus Infection

Model systems and an increased sensitivity of techniques in molecular biology have provided some insight into the functional aspects of papillomaviruses. In describing the data on DNA persistence in nonpermissive tumor cells and on gene expression we will follow the evolution of biological information from DNA via RNA to protein.

6.1 Physical State of Virus DNA in Tumor Cells

Integration of tumor virus nucleic acid into the host genome was long believed to be an essential step in cell transformation. Papillomaviruses are likely to break this dogma. Almost all systems investigated so far have given no evidence for integration into host cell sequences.

Only extrachromosomal virus DNA was detected in skin carcinomas of Ev patients (*Orth* et al. 1980; *Ostrow* et al. 1982; *Pfister* et al. 1983a) and in vulval and cervical carcinomas (*Green* et al. 1982).

In animal systems DNA persistence was studied with CRPV-induced benign and malignant lesions of domestic rabbits (*Stevens* and *Wettstein* 1979; *Wettstein* and *Stevens* 1980, 1982; *Favre* et al. 1982; *McVay* et al. 1982), with BPV 1- or 2-induced equine connective tissue tumors (*Lancaster* et al. 1979; *Amtmann* et al. 1980; *Lancaster* and *Olson* 1980; *Lancaster* 1981), and with BPV 1- or 2-induced hamster tumors (*Moar* et al. 1981a; *Pfister* et al. 1981a). Furthermore, cell cultures were established from the connective tissue tumors of cattle, horses (*Lancaster* and *Olson* 1980; *Lancaster* 1981), and hamsters (*Breitburd* et al. 1981), and from the transplantable Vx 7 carcinoma of rabbits (*McVay* et al. 1982); and in addition by the in vitro transformation of human keratinocytes (*LaPorta* and *Taichman* 1982), of C 127 mouse cells (*Lancaster* 1981; *Law* et al. 1981), and of bovine cells from the conjunctiva and the palate (*Moar* et al. 1981b).

The data on human bovine viruses can be summarized as follows:

1. The viral DNA persists with a high copy number of 25–500 genome equivalents per cell, which is in line with earlier findings with untyped BPV (*Lancaster* et al. 1976, 1977).

2. Circular DNA persists extrachromosomally and usually without any major deletions or rearrangements. This was principally shown by one or the other of the following experiments: (a) Supercoiled viral DNA could be purified by CsCl-ethidium bromide gradient centrifugation. (b) Southern blots of tumor DNA revealed virus-specific bands which comigrated with supercoiled, open circle, and linear viral DNA. (c) Cleavage of tumor DNA with restriction enzymes, which do not cut viral DNA, did not change the pattern of virus-specific bands in comparison to uncleaved DNA. (d) Single-cut enzymes of viral DNA converted all virus-specific DNA into full-length linear form. Only in one hamster tumor cell line were two additional bands observed after Bam HI and Eco RI digestion, which were tentatively interpreted as signals of integrated DNA (*Breitburd* et al. 1981). (e) Virus-specific cleavage patterns with various multicut restriction enzymes were identical with DNA taken from tumor cells and from virus particles.

Both in an Ev carcinoma and in an equine sarcoid viral DNA molecules were detected, bearing deletions of approximately 20% and 9%, respectively (*Ostrow* et al. 1982; *Amtmann* et al. 1980). As the deletion mutants appeared only together with wild-type DNA their biological activity cannot be evaluated at the moment.

3. High-molecular weight viral DNA was observed in small amounts in Ev carcinomas (*Ostrow* et al. 1982; *Pfister* et al. 1983a), in hamster tumors, and in transformed mouse cells (*Breitburd* et al. 1981; *Law* et al. 1981; *Pfister* et al. 1981a). These DNA molecules are not integrated by the above-mentioned criteria and are interpreted as concatemeric or catenated

structures. Digestion of Ev carcinoma DNA with nuclease S1, which converts supercoiled papillomavirus DNA into linear form, led to the formation of linear dimers (*Pfister* et al. 1983a), which would be expected in the case of concatemers. However, S1 treatment of DNA from BPV 1-transformed C 127 mouse cells destroyed the slow-migrating viral DNA, thus arguing for catenates (*Law* et al. 1981).

4. Sensitivity tests usually allow the conclusion that only less than 0.1–1 genome equivalent per cell could be integrated into cellular DNA. As far as studies with tumor biopsy material are concerned, one cannot exclude the possibility that integration of viral genomes happens in some small parts of the tumor. BPV 1-transformed C 127 mouse cells, however, were cloned, and the clones were shown to have less than 0.1 virus genome equivalent per cell integrated (*Law* et al. 1981). As these clones are able to induce tumors in nude mice it is clear that integration of a major part of the viral DNA is not essential for tumorigenicity in this system.

The situation is somewhat different with CRPV. A comparable amount of 10–100 copies of the viral genome is present per diploid cell of papilloma, primary, and metastatic carcinomas of the domestic rabbit. The majority of this DNA, however, behaved as a high-molecular weight complex when uncleaved tumor DNA was separated electrophoretically, and supercoiled CRPV DNA became only visible after prolonged exposure of Southern blots. Further analysis by two-dimensional gel electrophoresis, CsCl-propidium diiodide gradient centrifugation, partial digestion with restriction enzymes, and S1 nuclease treatment revealed episomal viral DNA circles of increasing size (*Wettstein* and *Stevens* 1982).

Single-cut enzymes of viral DNA led to full-length linear DNA, representing 99% of the total virus-specific DNA. In addition, faint bands were observed, which migrated slower or faster than linear CRPV DNA and could be interpreted as signals of integrated viral DNA. The pattern of the minor bands was reproducible among three different metastatic nodules of the same primary carcinoma but differed in a carcinoma of a second animal (*Wettstein* and *Stevens* 1980).

Favre et al. (1982) described extrachromosomal viral DNA of high molecular weight in the Vx 7 carcinoma also. Using a Vx 7 carcinoma which was independently transplanted for several generations, *McVay* et al. (1982) found exclusively integrated DNA in multiple copies. Eco RI cleavage patterns revealed five to six viral-cellular junction pieces. This observation obviously represents a rare exception, but CRPV seems to have an increased tendency for integration.

Nevertheless, extrachromosomal replication of viral DNA turned out to be a common characteristic of papillomaviruses. This ability does not exclude integration, however, as demonstrated by some examples.

Persisting virus DNA from both DNA (*Desrosiers* et al. 1979) and RNA tumor viruses (*Cohen* 1980) was described to be highly methylated. In contrast, analysis of BPV 1 DNA from hamster tumors with methylation-sensitive restriction enzymes Hpa II and Hha I gave no evidence for methylation (*Pfister* et al. 1981a). As the recognition sites of Hpa II and Hha I are distributed over the whole BPV 1 genome, this statement is fairly representative.

Partial methylation was observed with HPV 1 DNA, which was extracted from a single wart (*Danos* et al. 1980). One out of four Hpa II cleavage sites was methylated in about 40% of the DNA population. It will be interesting to see if this methylation plays a regulatory role in gene expression.

6.2 The Origin of Replication

The sequences of HPV 1 and BPV 1 show a region of approximately 1000 bp without significant open reading frames (*Danos* et al. 1982; *Chen* et al. 1982; see also Fig. 2). By analogy with polyoma-like viruses it is tempting to speculate that this region harbors the origin of DNA replication. Similar to the noncoding regions of SV 40 and BK, those genome segments reveal A + T-rich clusters, palindromic sequences, and direct and inverted repeats.

It should be pointed out, however, that a limited sequence homology between HPV 1 DNA and the origins of replication from SV 40, polyoma, and BK was identified in a complemetely different region of the HPV 1 genome (*Danos* et al. 1982). The sequence homology is about 3600 bp away from the region with the prominent signal structures. Therefore, functional data are necessary to define the origin of DNA replication of papillomaviruses.

6.3 Viral Transcription

RNA synthesis has been studied, so far, only with BPV 1 and CRPV. Viral transcripts are usually classified as early or late according to the temporal organization during the productive cycle. Early mRNAs are transcribed soon after infection and code for proteins, which are important for viral replication. With tumor viruses early proteins play a key role in the transformation of the host cell. Late mRNA transcription requires DNA synthesis, occurs at the end of the infection cycle, and leads to synthesis of structural proteins.

As has been pointed out already, no productive cell culture system exists for papillomaviruses, so that there is no experimental basis to discriminate between early and late genes. However, viral transcripts were analyzed from the virus-producing periphery of BPV 1-induced fibro-papillomas (*Amtmann* and *Sauer* 1982b) and compared to viral transcripts from the fibroma moiety of the wart, where no mature particles were detected. The latter transcripts were similar to those found in BPV 1-induced hamster tumors and in BPV 1-transformed mouse cells (*Amtmann* and *Sauer* 1982b; *Freese* et al. 1982; *Heilman* et al. 1982). In analogy with polyoma-like viruses and for convenience they will be referred to as "early transcripts." The mRNAs, which are specific for the wart periphery, are regarded as late. Similarly, CRPV transcripts in nonproductive papillomas and carcinomas are called early.

6.3.1 Early Transcripts

At least five early transcripts could be disclosed in the basal part of BPV 1-induced warts (*Amtmann* and *Sauer* 1982b). Two RNAs with 1100 and 1300 nucleotides, respectively, were by far the most prevalent ones. RNAs of the same size and equal prevalence were detected in transformed mouse cells and in hamster tumor cells (Table 6; *Freese* et al. 1982; *Heilman* et al. 1982). In the hamster system, however, it has not been possible, so far, to demonstrate two bands after gel electrophoresis, and it remains to be established if this is due to quantitative differences or to small size differences which make the bands melt together. Minor transcripts exist in the various systems, which differ considerably in size and quantity (Table 6). Some discrepancies may be due to experimental variation, but some may reflect altered viral gene expression in different host cells.

All early transcripts are polyadenylated (*Amtmann* and *Sauer* 1982b; *Freese* et al. 1982; *Heilman* et al. 1982). S1-mapping of the virus transcripts from C 127 mouse cells showed that they share a common 3'-end and extend to different 5'-termini (Fig. 4; *Heilman* et al. 1982). Comparable transcripts in bovine warts and hamster cells map in the same position as those from C 127 mouse cells (*Amtmann* and *Sauer* 1982b; *Freese* et al. 1982).

The RNAs were transcribed from the same DNA strand without evidence for internal splicing (*Heilman* et al. 1982). The presence of a remote 5'-leader sequence could not be ruled out, however, in those experiments; indeed it is suggested by Northern blot hybridization of viral RNA to subgenomic ^{32}P-labeled DNA fragments: RNAs with 1100, 1300, 1600, and 1800 nucleotides all gave a signal with the remote Sma I-Hpa I fragment (Fig. 4; *Amtmann* and *Sauer* 1982b).

BPV 1-specific RNA is only present in low quantity: 15–30 copies per C 127 mouse cell were estimated from dot blot intensities (*Heilman* et al. 1982), corresponding to 0.006%–0.01% of the total polyadenylated RNA pool in transformed cells.

Table 6. Early transcripts of bovine papillomavirus 1 and cottontail rabbit papillomavirus

	Virus					
	BPV 1				CRPV	
Cell/tumor system	Bovine wart	C 127 Mouse fibroblasts	C 57 mouse embryo fibroblasts	Hamster tumor	Hamster embryo-fibroblasts	Domestic rabbit papillomas/carcinomas
Gel system	Methyl-mercury	Formaldehyde	Methyl-mercury	Formaldehyde methyl-mercury	Methyl-mercury	Glyoxal
RNA classes	1.1 1.3 (1.6) (1.8) (2.9)	1.05 1.15 (1.7) (3.8) (4.05)	1.1 1.3 1.6 1.8	1.1–1.3 (1.6) (1.8)	1.1–1.3 1.6 1.8	1.3 1.3 2.0 2.0 (3.5) (3.5) (4.8) (4.8)
References	*Amtmann* and *Sauer* (1982b)	*Heilmann* et al. (1982)	*Amtmann* and *Sauer* (1982b)	*Amtmann* and *Sauer* (1982b); *Freese* et al. (1982)	*Amtmann* and *Sauer* (1982b)	*Nasseri* et al. (1982)

The most prominent transcripts in the individual systems are underlined. Minor RNA components are given in brackets.

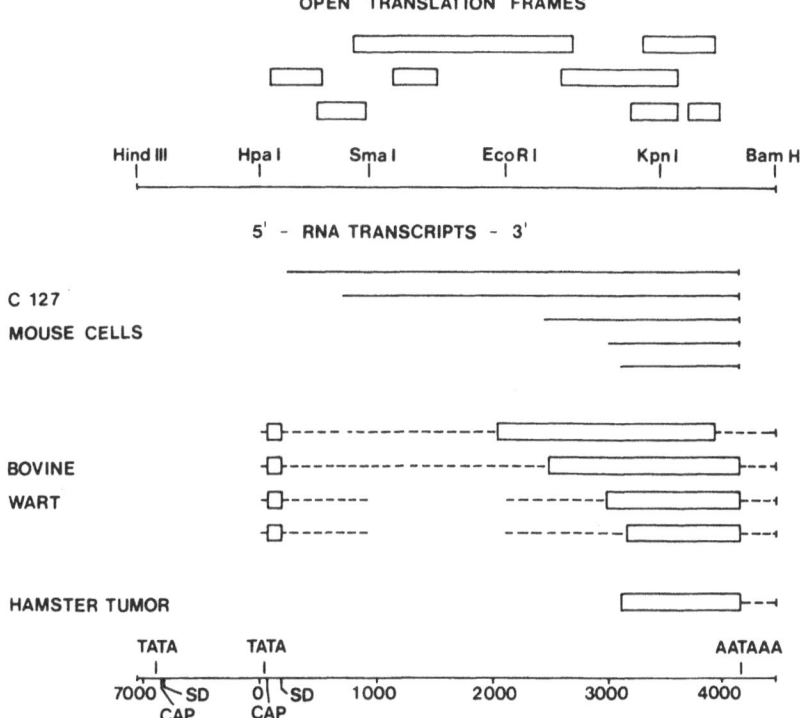

Fig. 4. BPV 1 RNA transcripts in transformed cells and in tumors, which do not produce mature particles or capsid antigen: C 127 mouse cells, fibroma part of the bovine wart, and hamster tumor. The transcribed genome segment is characterized by restriction enzyme cleavage sites *(top)* and nucleotide numbers *(bottom)*, as published by *Chen* et al. (1982). The open reading frames for proteins (see also Fig. 2) are given at the *top*. The location of transcripts in C 127 cells was defined by S1 mapping (*Heilman* et al. 1982). The approximate location of transcripts in bovine warts and hamster tumors, as determined by Northern blot hybridization to subgenomic DNA fragments, is indicated by *dotted lines*, the length of the transcripts by *boxes* (*Amtmann* and *Sauer* 1982b; *Freese* et al. 1982). If possible, the box position was aligned with the transcripts in C 127 cells. The location of promotor consensus sequences *(TATA)*, potential capping *(CAP)* and splice donor *(SD)* sites, as well as poly adenylation signals *(AATAAA)*, is indicated *below*. A possible leader sequence, as derived from the transcription control signals, is tentatively drawn at the 5'-end of the transcripts in bovine warts. These transcripts were shown to hybridize with the Hpa I-Sma I DNA fragment

DNA–RNA reassociation analysis with RNA from CRPV-induced skin carcinomas revealed virus-specific transcription covering 6%–12% of the viral genome (*Wettstein* and *Stevens* 1981; *McVay* et al. 1982). This is in the same size range as in the BPV 1 hamster tumor system. Northern blots of RNA from papillomas and carcinomas of the domestic rabbit revealed two major transcripts with 1300 and 2000 nucleotides, respectively (*Nasseri* et al. 1982). Two larger RNAs (3500 and 4800 nucleotides) were only detected with total cellular RNA but not with cytoplasmic RNA, suggesting that they are intranuclear precursors. The major transcripts were

mapped by Northern blot hybridization to subgenomic CRPV DNA fragments. They are colinear in large parts and both should be spliced, as deduced from hybridization to two noncontiguous DNA fragments. It is
interesting to note that the relative amount of the two RNAs varied between papillomas and carcinomas, the smaller RNA being more prominent
in carcinomas.

6.3.2 Late Transcripts

In the keratinized periphery of BPV 1-induced warts, where mature viruses
appear, a RNA species with 2000 nucleotides was disclosed in addition to
early transcripts (*Amtmann* and *Sauer* 1982b). This RNA is transcribed
from the same strand as the early RNAs and maps within the small Bam
HI-Hind III fragment of BPV 1 DNA (see Fig. 2). RNA from this region
was shown to code for the major structural protein of BPV 1 in in vitro
translation assays (*Heilman* and *Howley,* personal communication).

6.3.3 Reading Frames and Transcription Control Signals

With all papillomaviruses sequenced so far only one DNA strand has significantly large open reading frames which are likely to code for proteins.
This correlates with finding only one DNA strand to be transcribed. It suggests that the whole genetic information is coded by one strand, which is
in contrast to polyoma-type viruses, where early and late genes are transcribed from different strands (*Tooze* 1980).

   In the case of BPV 1, there are two TATAAA sequences which are
characteristic for eukaryotic promoters, just upstream of the early region
(see Fig. 4; *Chen* et al. 1982). A potential cap site for RNA (sequence
ACA) follows 31 nucleotides after the second TATA box. A short run of
125 nucleotides to a potential splice donor sequence (AGGTGCAT) could
give rise to a short leader sequence. Interestingly this part lies within the
very same Sma I-Hpa I fragment (see Fig. 4) which hybridized to four
early transcripts of BPV 1 (see Sect. 6.3.1). The first TATA box is also
followed by a cap site and a potential splice donor sequence, which would
give rise to a leader of only six nucleotides. No further promoter consensus
sequences appear within the early region.

   Functional assays for promoter activity led to congruent results
(*Campo* et al. 1983). BPV 1 Hae III DNA fragments were cloned into
a plasmid carrying the Herpes simplex virus type 1 thymidine kinase
(TK) gene deprived of its own promoter. The two Hae III fragments,
which harbored one of the above-mentioned promoter blocks each, were
able to stimulate TK gene expression. The fragments were tested in both
orientations, and their activity was significantly higher when oriented 5'-
to 3'- to the direction of transcription of the TK gene. Two additional

Hae III fragments from the 3'-end of the early region revealed enhancer properties (see below). All the other fragments were unable to replace the TK gene promoter.

A polyadenylation recognition sequence is found close to the 3'-end of the early transcripts.

Taken together, the data suggest that the early region of BPV 1 represents a single transcriptional unit where different transcripts are generated by differential splicing. Potential splice acceptor sites exist in the regions of the 5'-ends of all RNA transcripts from transformed C 127 mouse cells (*Heilman* et al. 1982; *Chen* et al. 1982).

A transcription enhancer was described upstream of the early region of SV 40, which activates transcription irrespective of polarity and distance to a promoter (*Gruss* et al. 1981; *Benoist* and *Chambon* 1980). *Lusky* et al. (1982) were able to substitute for the SV 40 enhancer with a BPV 1 sequence from the 3'-end of the early region. The role of this sequence in BPV 1 transcription remains to be established.

In the late region there are several potential promoters: In view of the reading frames, a TATATA sequence in front of L 2 and a TATAAA sequence in front of L 1 are most interesting. Two polyadenlyation recognition sites map close to the 3'-end of the late transcript (*Amtmann* and *Sauer* 1982b).

The distribution of transcription control signals is very similar in the case of HPV 1 (*Danos* et al. 1982). In contrast to BPV 1 there is a beautiful promoter consensus sequence at the beginning of the presumable late region: a CCAAT box followed by the TATAAT box after 39 nucleotides. CCAAT and TATAAAT boxes were also found in front of the late region of HPV 8 (*Fuchs* and *Pfister,* unpublished work). It is interesting to note that five noncontiguous regions at the 3'-end of the HPV 1 early region show partial homology with pieces of the SV 40 transcription enhancer (*Danos* et al. 1982). These sequences may reveal similar enhancer properties as the corresponding segment of the BPV 1 genome.

6.4 Nonstructural Viral Proteins

It has not been possible to disclose nonstructural proteins of papillomaviruses in any system up to now. Some indirect evidence exists in the cases of CRPV and BPV 1 or 2, but the virus specificity of these antigens or proteins has not been proven.

Both nuclear (*Ishimoto* et al. 1970) and membrane fluorescence (*Ishimoto* and *Ito* 1969) were detected in indirect immunofluorescence tests with cells from CRPV-induced papillomas of the cottontail rabbit and sera from papilloma-bearing rabbits. In most cases only less than 5% of the cells were positive. Cells from domestic rabbit papillomas were either negative

(*Orth* and *Croissant* 1968) or showed a similar pattern to that of cotton-tail cells with sera from young tumor-bearing animals (*Seto* et al. 1977). A tumor-specific transplantation antigen of CRPV-induced papilloma and carcinoma cells was suggested by cell-mediated immune reactions in vitro (*Hellström* et al. 1969).

Some contradictory data were reported on antigens of BPV-transformed cells. Cytoplasmic fluorescence was observed in 20% of BPV-transformed hamster and mouse cells by using sera from tumor-bearing hamsters (*Geraldes* 1970). In an independent assay no fluorescence was detected with hamster sera in cells derived from BPV-induced tumors from cattle, horses, and hamsters (*Barthold* and *Olson* 1978); however, sera from cattle and horses bearing BPV-induced fibromas reacted with cell membranes of the tumor cells. No intracellular antigens were found in this study. Five polypeptides were reproducibly precipitated from extracts of BPV 1-induced hamster tumor cell lines by sera from tumor-bearing hamsters and pikas (*Breitburd* et al. 1981). The molecular weights were 190 000, 125 000, 59 000, 33 000, and 30 000. Specificity controls were carried out with SV 40-transformed cell lines and a normal hamster embryo cell strain, as well as with sera from hamsters bearing SV 40-induced tumors. A number of cross-reactions indicate that these polypeptides are most likely transformation-related proteins, but not necessarily papillomavirus specific.

Antigens with similar properties were detected in human warts: A nuclear and a cell surface antigen were purified from wart homogenates by immunoadsorbent chromatography (*Pass* and *Marcus* 1973). The antigens were present in warts, squamous cell carcinomas, fetal skin and psoriatic epidermis, and even in concentrated extracts of normal skin.

One reasonable approach to early proteins of papillomaviruses should be in vitro translation of the virus-specific RNA from BPV 1-transformed cells. Whereas late mRNA has been successfully translated into the capsid protein, however, no early proteins have been translated so far (*Heilman*, personal communication; *Fink* and *Pfister*, unpublished work).

In spite of frustrating experiences in the search for early proteins, there can be little doubt that such proteins exist. A comparison of nucleotide sequences and theoretical amino acid sequences of HPV 1, HPV 6, and BPV 1 clearly shows that protein structures are highly conserved within reading frame E 1. Nucleotides are often exchanged in the third codon positions, where an exchange does not affect the amino acid coded for (Fig. 5). This can only be explained by an evolutionary pressure on the protein level. From the limited viral RNA synthesis in tumors and in transformed cells one may expect only low amounts of protein. Furthermore, the proteins may be weakly antigenic. Both assumptions could account for the negative results obtained so far.

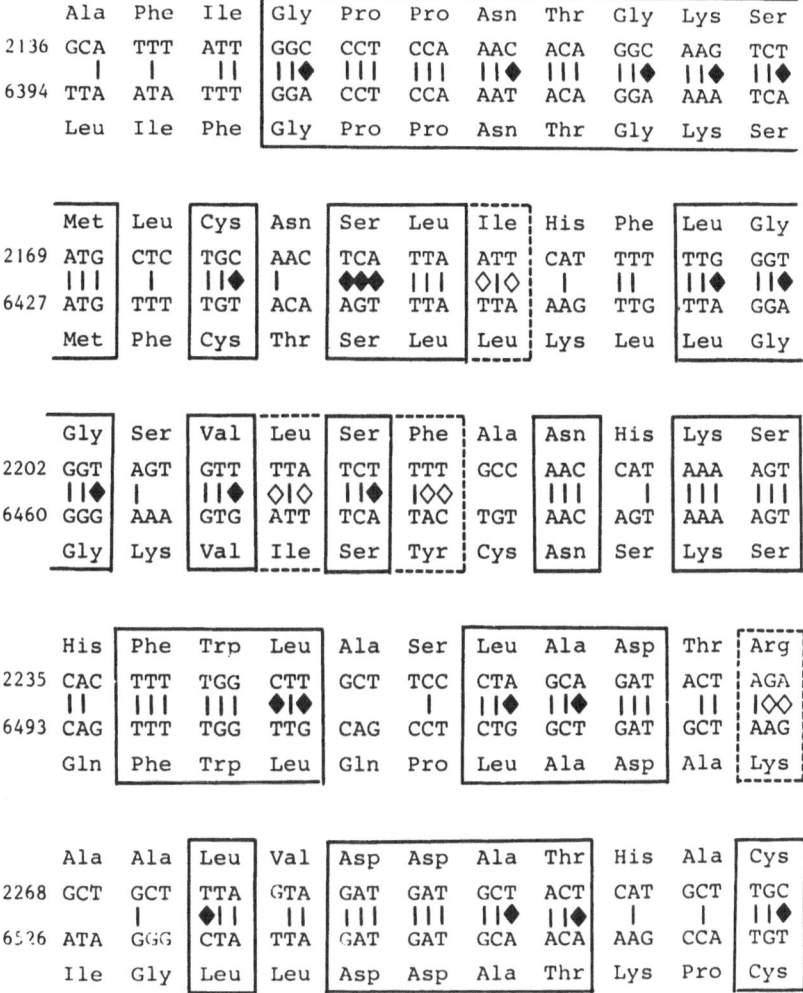

Fig. 5. Partial nucleotide and amino acid sequences within reading frame E 1 of BPV 1a *(upper)* and HPV 1a *(lower)*. The nucleotide numbers correspond to those of the published sequences (*Chen* et al. 1982; *Danos* et al. 1982). Conserved amino acids are *boxed;* chemically similar amino acids are framed by *dotted lines.* Identical nucleotides are connected by a *line.* Nucleotide exchanges which do not affect the amino acid coded for are displayed by *closed rhombs.* Nucleotide exchanges which lead to chemically similar amino acids are indicated by *open rhombs*

6.5 Mechanism of Transformation and Tumor Induction

New insights into the genome structure and the gene expression of papillomaviruses have shown that they are only distantly related with polyoma-like viruses. Therefore, it is possible that transformation and tumor induction by papillomaviruses differs considerably from the well-known tumor virus system.

Several attempts have been made to define the transforming principle of papillomaviruses. Transfection studies with BPV 1 DNA and mouse cells showed that the entire genome is not needed for transformation to growth in soft agar and tumorigenicity in the nude mouse (*Lowy* et al. 1980). Transfection with subgenomic fragments confined the essential region to the 3.4×10^6 Bam HI-Hind III fragment, representing the early region of the viral chromosome (see Fig. 4). Transfection efficiency varied between 13% and 35% of the values obtained with full-length linear DNA; this may be due to difficulties in establishing the DNA in the cells. As observed with full-length DNA, subgenomic fragments have to recircularize after transfection, but, in contrast to the complete genome, where this usually works properly, circularization of fragments is accompanied by the acquisition of additional sequences, duplications, or rearrangements (*Law* et al. 1981).

Deletion mutants were constructed in vitro, allowing a more precise characterization of the region essential for transformation (*Lowy* et al., personal communication; *Howley* et al., personal communication). The 5'-end of the early region seems to provide only the promoter (see Fig. 4). This was elegantly shown by ligation the 1.4×10^6 Eco RI-Bam HI fragment to the large terminal repeat of a retrovirus, thus substituting for the promoter function and getting a transforming DNA molecule (*Nakabayashi, Chattopadhyay* and *Lowy,* personal communication). According to the present state of knowledge, the transforming ability resides between nucleotides 2405 and 4450 (Bam HI), i.e., within 25% of the viral genome.

As has been already discussed in detail (see Sect. 6.1), there is no evidence for integration of BPV DNA. Therefore, there is no molecular basis for a possible cell transformation by integration of a viral promoter in front of a cellular oncogene, as described for retroviruses (*Hayward* et al. 1981).

The persistence of BPV 1 DNA alone does not seem to be sufficient for transformation. Bovine fetal thyroid cells are capable of supporting continuous BPV 1 DNA replication without detectable viral gene expression and retain a normal phenotype (*Amtmann* and *Sauer* 1982a). Treatment of infected cells with the tumor promoter 12-*O*-tetradecanoyl-phorbol-13-acetate (TPA) induced viral transcription and also cell transformation in terms of altered morphology, reduced serum requirement, and increased saturation density. This indicates that viral gene expression is necessary for transformation.

Induction of viral transcription by TPA has similar consequences in other systems. BPV 1 DNA did not replicate in embryonic fibroblasts of DBA mice and was lost after a few cell passages (*Amtmann* and *Sauer* 1982a). The DNA was not transcribed and the cells were phenotypically normal. TPA treatment induced several species of BPV 1 RNA, and viral DNA concentration increased considerably as early as 18 h after treatment. After stimulation the BPV 1 DNA continued to replicate in the absence of TPA, transcriptional activity proceeded, and the cells appeared transformed. Tumor promoter activity may thus enable papillomaviruses to establish themselves in cells which are, a priori, not susceptible at all.

TPA also activates papillomavirus DNA in vivo. Treatment of the multimammate mouse drastically increased the number of persisting viral genomes per cell (see also Sect. 4.3), and this event probably depended on stimulation of gene expression (*Amtmann* and *Sauer*, personal communication). It is of special interest that after TPA treatment tumors arose significantly earlier than the spontaneous tumors of the control animals.

In summary, the above-mentioned experiments indicate that cell transformation in vitro and tumor induction in vivo apparently depend on viral gene expression, at least on RNA transcription. Up to the present, no temperature-sensitive mutants which would be in favor of a transforming protein have been found to exist.

Nothing is known about the molecular mechanisms that might be involved in malignant conversion of papillomas. Comparative analysis of individual virus types is still in its infancy and has not yet revealed relevant differences between the types of viruses which were found to be associated with carcinomas and those which were detected in benign lesions.

Chromosomal abnormalities are well known to be associated with malignancies of humans and animals and are supposed to play an important role in carcinogenesis. As a possible mechanism of tumor induction it was suggested that virally induced cell proliferation may enhance the risk for mutations, which could affect the balance between tumor virus activity and host cell control (*Zur Hausen* 1977b). Some preliminary observations with papillomaviruses are in line with this hypothesis. Histological studies on biopsies from women with severe cervical dysplasia and condylomas, and with dysplasias alone, revealed significantly more mitotic abnormalities in the condylomatous lesions, which is consistent with viral alterations of mitosis (*Boon* and *Fox* 1981). BPV 1 induced fibrosarcomas of Syrian hamsters were tested for chromosomal abnormalities and all showed abnormal karyotypes (*Gamperl* et al. 1983). Chromosomes 1, 4, and 15 were most frequently stricken. These chromosomes are also affected in many hamster tumors induced by other viruses or by chemical carcinogens (*Pathak* et al. 1981; *Dipaolo* and *Popescu* 1979); this implies a convergent action of different oncogenic agents. Large-scale screening experiments will be necessary to substantiate these data.

6.6 BPV 1 as Cloning Vector

Some properties of BPV 1 suggested the usefulness of BPV DNA as a
eukaryotic cloning vector, thus opening a new field of papillomavirus
research (*Sarver* et al. 1981). As described in Sect. 6.5, a subgenomic fragment is able to transform susceptible mouse cells, and the DNA persists
extrachromosomally with multiple copies. This ensures both the amplification and physical integrity of the cloned DNA. The transformed character
of the cells provides a further amplification step as a result of rapid cell
division and offers a selection marker for those cells that have incorporated
the foreign DNA segment.

In a first assay the rat preproinsulin gene I was linked to the transforming region of BPV 1 (*Sarver* et al. 1981) and transfected to C 127
mouse cells. A correctly spliced preproinsulin transcript could be detected
in transformed cells, and rat proinsulin protein was demonstrated in the
culture medium by radioimmunassay and immunoprecipitation.

In an attempt to obtain independence from the transformed phenotype as selective marker for transfection, the transforming region was
linked to a dominant selective marker: the *Escherichia coli* xanthine
guanine phosphoribosyltransferase (XGPT) gene (*Law* et al. 1982). The
new vector transferred two phenotypes, which could be selected for
separately. One difficulty encountered with this vector was the high frequency of rearrangements of the input DNA, the reason for which is not
known at present.

In many instances it is extremely helpful when cloned genes can be
shuttled between mammalian cells and bacteria. With BPV 1 as eukaryotic
vector moiety a shuttle vector was constructed by ligation with a pBR 322
deletion derivative (*Binetruy* et al. 1982; *DiMaio* et al. 1982; *Sarver* et al.
1982). The pML 2 variant of pBR 322 lacks the sequence that inhibits replication in eukaryotic cells (*Lusky* and *Botchan* 1981). The BPV 1-pML 2
hybrid molecule replicates as a stable, unintegrated, multicopy plasmid in
mouse C 127 cells and in FR 3T3 rat fibroblasts (*Binetruy* et al. 1982)
and can be rescued in bacteria.

In the meantime, human β-globin gene (*DiMaio* et al. 1982), human
interferon β1 gene (*Mitrani-Rosenbaum, Maroteaux, Mory, Revel,* and
Howley, personal communication), and hepatitis B virus surface antigen
(*Stratowa, Wang, Schäfer-Ridder,* and *Hofschneider,* personal communication) were cloned in the BPV 1 vector and were transferred to mouse cells.
They are all transcribed in the eukaryotic cells. Interferon was produced
constitutively at low levels and responded to induction with inactivated
Newcastle disease virus or poly IC. Hepatitis B virus surface antigen was
stably produced in high amounts. This indicates that BPV 1 offers a successful new vector system for molecular biology studies in eukaryotes.

7 Immune Response

Man reacts to papillomavirus infection with humoral and cellular immune mechanisms against virus particles and tumor tissue (*Thivolet* and *Viac* 1978). This is reflected by the facts that papillomavirus infections are mainly acquired by children or early adolescents, that many warts regress spontaneously after variable periods, that rejection is systemic in the case of multiple warts, and that the hosts are left immune to reinfection later in life. The special importance of cell-mediated immunity becomes clear from the following observations:

1. Patients treated with immunosuppressive drugs, for example after kidney transplantation, often suffer from disseminated, nonregressing warts (*Spencer* and *Anderson* 1970; *Starzl* et al. 1970; *Koranda* et al. 1974).
2. Patients with cell-mediated immune deficiencies secondary to Hodgkin's disease or chronic lymphatic leukemia are more prone to papillomavirus infection than patients with humoral immune deficiencies (*Perry* and *Harman* 1974; *Morison* 1975a; *Reid* et al. 1976; *Ward* et al. 1977).

Nevertheless, the roles of the different immune mechanisms in wart regression and prevention are still not well understood. This is certainly at least partially due to the use of pooled virus preparations as antigen source in earlier assays, which probably obscured type-specific reactions. Many immunological experiments will have to be repeated with characterized viral antigens.

For further discussion it may be useful to start with a brief summary on the immunology of Shope papillomavirus infection. This system has provided a lot of data on the immune response to a clearly defined virus and offers a solid basis for the discussion of more recent results on HPV.

7.1 Shope Papillomavirus

Rabbits respond to papillomavirus infection by generating neutralizing antibodies which protect from reinfection (*Kidd* et al. 1936). The antibodies are demonstrable in animals with both regressing and persisting tumors (*Kidd* 1938; *Seto* et al. 1977), indicating that they do not cause regression. The importance of other mechanisms in this respect was clearly shown in reinfection experiments, where virus inactivation by antibodies was avoided either by infection of autologous skin in vitro followed by transplantation (*Kreider* 1963) or by transfection with purified viral nucleic acid (*Evans* and *Ito* 1966): Papillomas were only induced in rabbits with persisting papillomas and not in regressor animals. Immunity lasted for at least 5 months after regression and was independent of the life span of the first tumor (*Evans* and *Ito* 1966). This type of immunity is likely to be directed against tumor antigens, as vaccination with tumor tissue increased the frequency of regression from 25% (spontaneous) to 50%–90% (*Evans* et al. 1962). Mononuclear cell infiltrates in regressing papillomas point to cellular immune mechanisms being the active principle in papilloma regression (*Kreider* 1980). Leukocytes were most concentrated at the epithelial basement membrane, whereas reduction of cell proliferation was most obvious in the spiny layer; this may indicate the activity of lymphokine-like substances. Cellular immune mechanisms were also demonstrated by in vitro studies (*Hellström* et al. 1969): Lymph node cells from rabbits in which papillomas had spontaneously regressed inhibited colony formation of papilloma- and carcinoma-derived cells. Surprisingly this was also achieved by lymph node cells from nonregressor animals; however, in contrast to regressor rabbits, nonregressor sera abrogate the inhibitory effect. This implies a cell-mediated immune response to the surface of tumor cells and a sensitivity of this reaction to the humoral factors which play an important role in preventing regression. Additional immune reactions were disclosed after CRPV infection of newborn domestic rabbits (*Seto* et al. 1977). Fifteen weeks after infection delayed-type hypersensitivity to CRPV was detected in three regressors and in five out of eight persisters. Besides neutralizing antibodies, antibodies appeared against nuclear and cytoplasmic papilloma cell antigens in both persisters and regressors, thus giving no direct evidence for a role in wart regression.

The overall effect of the immune system in CRPV infection is beautifully demonstrated by the results of immunity impairment by methylprednisolone (*McMichael* 1967). Steroid treatment did not influence latency period, papilloma growth rate, virus concentration, and malignant conversion of papillomas. However, treated animals suffered from "secondary" papillomas, arising at sites not directly inoculated with virus, and the tumors hardly regressed. Only 2.5% of steroid-treated animals showed regression, whereas 47% of the tumors of control animals regressed. Furthermore, methylprednisolone led to increased susceptibility to reinfection by CRPV.

7.2 Human Papillomaviruses

7.2.1 *Humoral Immunity*

7.2.1.1 *Response to Virus Structural Proteins*
Regarding the heterogeneity of human papillomaviruses, most of the data on antibody response to HPV must be interpreted carefully as they were obtained with antigen from pooled warts. As far as plantar wart patients

were tested, the conclusions will mainly hold true for HPV 1 because of the prevalence of this virus type in plantar warts and the relatively high particle yield with HPV 1.

Antibodies reacting with HPV particles were first demonstrated by electron microscopy by *Almeida* and *Goffe* (1965). They are directed towards the major structural protein of the viral capsid (*Pass* and *Maizel* 1973). Follow-up studies with individual patients showed that it may take several months to develop antibodies after the onset of warts (*Cubie* 1972). Of former plantar wart patients 70% were still antibody positive 2 years after healing, and so were 20%–30% up to 9 years after infection. Countercurrent immunoelectrophoresis was used as antibody detection test for this study. According to a monospecific solid-phase radioimmunoassay, the percentage of anti-HPV 1 antibody-positive people in the unselected populaton rises with age to reach a maximum of about 50% at 20 years and gradually declines afterwards to an average value of 35% (*Pfister* and *Zur Hausen* 1978b). It is interesting to note that the incidence of HPV 1-induced warts drops sharply in the 20–30 years age group, and one might speculate about a protective role of the antibodies.

HPV 2, HPV 3, HPV 5, and HPV 8 were also shown to elicit antibody response by using monospecific antigens for immunodiffusion, immunofluorescence, or immune electron microscopy tests (*Jablonska* et al. 1979, 1980; *Pfister* et al. 1979a, 1981b). Eight out of 11 Ev patients with persistent warts had antibodies against one or more of the above-mentioned HPV types. Butchers, who have a high risk of papillomavirus infection, showed specific HPV antibodies in about 50% (25/51) in contrast to other adults, of whom 85% were negative in immunodiffusion tests (*Jablonska* et al. 1980). A limited seroepidemiology of HPV 8 by immune electron microscopy revealed type-specific antibodies at low titers in ten out of 100 healthy adults (*Pfister* et al. 1981b). This contrasts the fact that HPV 8-induced lesions were only observed in Ev patients and indicates a rather wide distribution of HPV 8 with subclinical infections.

The anti-HPV titers are usually low, which points to a very limited antigenic stimulation of wart carriers. Existing antibody titers slowly increase during the disease (*Pyrhönen* 1978). IgM antibodies appear first and predominate for long periods without detectable influence on tumor growth. At the onset of regression all patients had virus-specific IgM antibodies when examined with an indirect immunofluorescence test (*Matthews* and *Shirodaria* 1973; *Shirodaria* and *Matthews* 1975). In contrast, only 12% of the patients with nonregressing warts were IgM positive. Close to regression, IgG antibodies with complement-fixing activity come up (*Ogilvie* 1970; *Genner* 1971; *Pyrhönen* and *Penttinen* 1972; *Pyrhönen* and *Johannson* 1975). They were taken as signs of good prognosis and were discussed as playing a role in wart regression. *Brodersen* and *Genner* (1973), however, found no significant differences between regressors and persisters when testing for complement-fixing antibodies. On the other hand, a renal allograft recipient suffering from HPV 3-induced flat warts showed anti-HPV 3 antibodies of the IgG class with a fairly good titer of 1:80 in an indirect immunofluorescence test without any signs of regression. Furthermore, warts recurred after surgical removal (*Pfister* et al. 1979a). These findings indicate that anti-HPV IgG antibodies may well be a frequent attendant symptom of regression and have prognostic value, but they are not likely to be essential or sufficient for regression.

7.2.1.2 Response to Tumor Cell Proteins

Wart patients develop IgM and rarely IgG antibodies against tumor cell-specific antigens (*Matthews* and *Shirodaria* 1973; *Shirodaria* and *Matthews* 1975; *Viac* et al. 1978). This activity cannot be removed by adsorption of sera with virus particles and could theoretically be directed against non-structural viral proteins or newly induced cellular proteins. The antigens are located within the cytoplasm of some infected cells, and no label was detected at the cytoplasma membrane or in the intercellular space (*Viac* et al. 1978). At regression, 83% of the patients revealed such antibodies, in contrast to 46% of the patients with persisting warts (*Matthews* and *Shirodaria* 1973). This increase associated with regression could be explained by a more intimate contact with the intracellular antigens during wart involution.

7.2.2 Cell-Mediated Immunity

Epidemiological arguments in favour of a central role of cell-mediated immunity in wart rejection have already been mentioned (see page 157). Systematic investigations of nonspecific cell-mediate immunity in wart patients and on the histology of regressing warts have given further support to this theory.

7.2.2.1 Nonspecific Cell-Mediated Immunity

To test for a general defect in the cell-mediated immunity of wart patients the following parameters and tests were used: frequency and index of 1-nitro-2, 4-dichlorobenzene (DNCB) sensitization, percentage of E-rosette-forming (ERF) lymphocytes, lymphocyte transformation (LT), and leukocyte migration inhibition (LMI). Using LT, wart patients proved to be less responsive than control groups, and the reactivity was worse in the case of patients with long-lasting warts (*Morison* 1975b). The number of T cells was decreased in untreated wart patients (*Chretien* et al. 1978). Patients with different types of warts were studied separately and a rather heterogeneous picture was found with DNCB sensitization, ERF, and LT tests

(*Glinski* et al. 1976, 1981; *Obalek* et al. 1980). Whereas cell-mediated immunity was well preserved in the case of HPV 1-, HPV 4-, and HPV 7-induced skin warts, it was considerably disturbed in patients with HPV 2- and HPV 3-induced lesions, and severely impaired in Ev patients suffering from HPV 5, 8, or 9 infections. Depressed cell-mediated immunity of Ev patients was also described by *Prawer* et al. (1977) and by *Kienzler* et al. (1979). This contrasts the well-preserved humoral immunity of Ev patients with a number of virus-type-specific responses (*Jablonska* et al. 1980). Patients with genital warts were either inconspicuous (*Obalek* et al. 1980) or their lymphocytes were less responsive to both T cell and B cell mitogens (*Seski* et al. 1977). These observations may bascially lead to two conclusions:

1. The susceptibility to infections with different types of HPV is related to the grade of cell-mediated immunity depression (*Obalek* et al. 1980).
2. Infection with certain types of HPV leads to depression of cell-mediated immunity.

Iatrogen immunosuppression in connection with organ transplantation provided material for a prospective point of view and showed that multiple flat warts induced by HPV 3 are by far the most common type of lesion in these patients (*Morrison* 1975a; *Pfister* et al. 1979a). Quite recently, HPV 5-induced lesions were also detected (*Lutzner* et al. 1980). These findings indicate that defects in cell-mediated immunity do indeed predispose for infections with certain HPV types, according to hypothesis 1. On the other hand, cell-mediated immunity improved considerably in two cases of long-standing warts *after* surgical removal or spontaneous regression (*Reid* et al. 1976; *Jablonska* et al. 1982), implying that an already impaired immunity may be further depressed by some HPV types.

There is one interesting report on a soluble factor from long-persisting warts which is able to block the local expression of cellular immunity without affecting systematic responsiveness (*Freed* and *Eyres* 1979). The wart extract blocked LMI by purified protein derivative (PPD) in vitro, and there was no reactivity to DNCB in vivo when the drug was applied to the warts, whereas successful sensitization could be demonstrated on unaffected skin. This is the only direct evidence for local immune modulation by recalcitrant warts.

As described for the CRPV system (*McMichael* 1967), malignant conversion does not seem to depend on the immune status of the patient (*Glinski* et al. 1981). Cell-mediated immunity was impaired to the same extent in 13 Ev cases with lesions induced by HPV 3, HPV 5, or HPV 3 and 5, whereas only seven patients infected with HPV 5 developed Bowen's carcinomas. These findings indicate that malignancy depends more on the cancerogenic potential of the virus than on the extent of T-cell defect.

7.2.2.2 Specific Cell-Mediated Immunity

Specific tests were carried out with papillomavirus particles (*Lee* and *Eisinger* 1976; *Viac* et al. 1977a, b) or wart extracts as antigen (*Morison* 1975c; *Lee* and *Eisinger* 1976). Unfortunately, the virus type was not determined in these studies, but the antigen source suggests that it was mainly HPV 1 and 2. In some experiments homologous wart extracts were used for the test of individual patients. Most people with warts of less than 1 year's duration reacted both in vivo with delayed hypersensitivity (*Viac* et al. 1977a, b) and in vitro in LT or LMI tests (*Morison* 1975c; *Lee* and *Eisinger* 1976). Weak or nonexistent specific immune response correlated with persistence of warts (*Viac* et al. 1977a). The immunity is short-lived, and all persons had lost their positive in vitro response within 10–13 weeks after regression (*Morison* 1975c); however, this may be a question of test sensitivity.

7.2.2.3 Observations on Regressing Warts

Sensitization of human tissues with DNCB is known to induce local cell-mediated reactions and was successfully used to induce regressions of warts (*Greenberg* et al. 1973); indeed, regression of flat warts shows many characteristics of a cell-mediated immune reaction. The lesions spontaneously become red with itching and then disappear completely (*Tagami* et al. 1977; *Berman* and *Winkelmann* 1977). Histological examination revealed an intense mononuclear cell infiltrate in the dermis, accompanied by epidermal spongiosis and cell necrosis. The lymphocytes were mainly restricted to the basal layer and were not detected in the degenerating wart tissue (*Takigawa* et al. 1977). The authors described involution as rejection from the level of the basal cell layer, followed by recovery processes as in wound healing. Recent studies on flat warts at the earliest stage of involution disclosed macrophage attacks on degenerating epidermal cells (*Oguchi* et al. 1981b). Cell membranes frequently disappeared and macrophages invaded the epidermal cells. Macrophages also lose their membrane and pour out. In areas with activated macrophages Langerhans cells ultrastructurally showed signs of enhanced cellular activity (*Oguchi* et al. 1981a).

It is interesting to note that no inflammation was observed in the first-described case of regression of HPV 3-induced flat warts in an Ev patient (*Jablonska* et al. 1982). The lesions regressed slowly and progressively after two pregnancies, and this might imply a role for fetal antigens and/or other immunological factors related to pregnancy.

With plantar and common warts no cellular infiltrate was originally observed during regression (*Brodersen* and *Genner* 1973; *Matthews* and *Shirodaria* 1973). More recently, however, clinically apparent inflammation and mononuclear cell infiltrates have been described (*Berman* and *Winkelmann* 1980; *Berman* et al. 1982). An inflammation as the result of bacterial superinfection was unlikely, as clinically inapparent warts from distant sites of the same patient also showed infiltrates at that time. The warts finally turned black as the result of thrombosis of the blood vessels and regressed.

8 Diagnosis

HPV-induced warts are usually diagnosed by the clinician without any doubt, and the demonstration of virus particles is not necessary for confirmation. As outlined above (Sect. 7.2.1), the antibody response of the patients is weak and slow, so that serology does not provide much information either. For both these reasons virology has hardly played any role in wart diagnosis.

Two aspects which have become clear during the past few years have changed this situation. On the one hand, papillomaviruses have been detected in lesions which were not regarded as HPV induced and which are indeed difficult to identify as HPV induced by clinical criteria alone. On the other hand, malignant conversion of HPV-induced tumors seems to correlate with infection by certain virus types. If this holds true, virus classification will be extremely important for prognosis and will be the basis for appropriate prevention.

8.1 Test for HPV Etiology

To recap, there are HPV-induced pityriasis versicolor-like lesions, which are very characteristic in Ev patients but may be easily misdiagnosed if they are dispersed, for example in immunosuppressed patients, and there are cervical dysplasias, a number of which can be identified as condylomata plana by virological methods. Histology provides a first clue for diagnosis. Virology offers two routine tests with biopsy material for a final proof:

1. Demonstration of viral antigens in thin sections by means of a group-specific antiserum using the peroxidase-antiperoxidase technique (see Sect. 3.3). The antiserum is commercially available, and the assay can be carried out with formalin-fixed, paraffin-embedded material. In cases of low antigen production, a number of sections must be screened for final evaluation.
2. Demonstration of viral DNA by hybridization under relaxed conditions (see Sect. 3.3). This test is more time consuming and requires native material but is very useful in cases of low antigen production.

It would be worthwhile applying these tests to other lesions which have not yet been identified as HPV induced. For example, *Syrjänen* (1980, 1982) described koilocytotic epithelial changes very similar to flat condylomas in close proximity to bronchial and esophageal squamous cell carcinomas.

8.2 Virus Classification

Monospecific antisera are available for only a few HPV types (see Sect. 3.1). Therefore, the method of choice for virus typing is hybridization of DNA from biopsy material to cloned viral reference DNAs. In view of cross-hybridization between various HPV types (see Sect. 3.1), additional parameters are essential for exact classification. They may be obtained from restriction enzyme cleavage patterns and/or by determining the amount of cross-hybridization by reassociation kinetics. Needless to say, the biopsy material must be native and not formalin fixed to allow DNA extraction and hybridization experiments.

9 Treatment of Warts

A recent review of the treatment of warts was made by *Bunney* (1982). Her monograph mainly covers surgical intervention, cryotherapy, topical treatment with salicylic acid, caustics or podophyllin, and immunotherapy. It is stressed that the effectiveness of different methods is difficult to evaluate. (This is reflected by the ongoing debate on the efficiency of the mystic charming of warts.) As *Bunney* points out, the response to treatment is clearly influenced by the type of wart (i.e., in many cases the type of HPV!), the number and duration of the warts, and the age and immune reactivity of the patient. One "success" or another may be attributable to the timely onset of spontaneous regression.

Besides treatment at home with a variety of salicylic acid preparations, cryotherapy is at present the most universally used technique.

The effects of aromatic retinoids and interferon have recently been studied on the virus level and these aspects will be covered here.

9.1 Treatment with Retinoids

Hypervitaminosis A was shown to inhibit the growth of CRPV-induced papillomas (*McMichael* 1965). Synthetic aromatic retinoids have been developed, which are not as toxic as hypervitaminosis A (*Mayer* et al. 1978) but inhibit both the induction and the development of CRPV-induced papillomas (*Ito* 1981). Ro 10-9359 (Tigason) treatment (200 mg/kg given intramuscularly twice a week) led to complete regression of well-established tumors in about 60% of the animals, and there was no regrowth. All tumors at least showed marked retardation of growth and a reduction in tumor volume. The drug did not affect CRPV DNA-containing carcinomas Vx 2 and Vx 7, but there was a significant reduction in the number and size of pulmonary metastases. The reactivity of papillomas and carcinomas correlates with the different levels of cellular retinoic acid-binding protein. The binding capacity of papillomatous tissue was about 15 times greater than that of carcinomas or normal skin (*Rattanapanone* et al. 1981).

Oral retinoid treatment of humans (1 mg/kg per day) proved to be very successful in the beginning. HPV 3-, 5-, and 8-induced lesions of Ev patients (*Lutzner* and *Blanchet-Bardon* 1980; *Jablonska* et al. 1981; *Nürnberger* et al. 1981) and multiple HPV 2-induced common warts (*Gross* et al. 1983) considerably improved after few weeks and many warts regressed. Neither viral antigens nor viral DNA could be detected in the remaining lesions, and there was no histological evidence for virus replication. This may be explained by the interference of retinoids with keratinization, which might in turn affect HPV replication.

Unfortunately, the primary response is not sustained: The warts tend to recur when the drug is withdrawn or even when the dosage is reduced. The viruses were characterized before and after treatment and turned out to represent the same subtype (*Gross* et al. 1983), which indicates that Tigason works primarily as a suppressive and does not necessarily eliminate the virus. This hypothesis is in line with the observation that BPV 1-transformed C 127 mouse cells are not cured of viral DNA by in vitro treatment with Tigason (*Gassenmaier* and *Pfister*, unpublished work).

Wart recurrences after therapy are a serious problem, in view of severe side effects such as dryness of mucosae, nasal bleeding, scaling of the skin, hairloss, and fragility of the nails, which prevent continuous therapy in a number of patients. These problems indicate that Tigason should not be recommended for wart treatment in general; however, estimating benefit and risk, it may be helpful in some cases of severe, generalized verrucosis.

9.2 Treatment with Interferon

Twenty-one patients with severe juvenile laryngeal papillomatosis were treated with leukocyte (alpha) interferon (*Haglund* et al. 1981; *Goepfert* et al. 1982). The tumors initially decreased in size or regressed completely. However, even while receiving interferon, half of the patients suffered from recurrences after 4—8 months (*Goepfert* et al. 1982). When treatment was discontinued, papillomas frequently recurred but decreased again after therapy was restarted (*Haglund* et al. 1981).

Fibroblast (beta) interferon turned out to be ineffective in the treatment of larnygeal papillomatosis (*Göbel* et al. 1981). It was administered intravenously, and its failure could have been due to its fast clearance.

In vitro, mouse L-cell interferon had a pronounced effect both on BPV 1 transformation of mouse cells and on transformed mouse cells (*Turek* et al. 1982). Cultivation of BPV 1-infected C 127 cells in the presence of 200 units of interferon per millilitre led to a reduction of focus number to 1/20 of the untreated control. Furthermore, interferon lowered the BPV 1 genome copy number of established transformed C 127 cells to one-third to one-eighth. Reverters to the nontransformed phenotype could be isolated from interferon-treated cultures; they no longer contained detectable BPV 1 sequences and had lost the capacity to form colonies in soft agar. Complete curing of the treated culture was not achieved after 60 cell divisions.

Clinical and experimental data are very much in agreement, suggesting that in many cases interferon will be extremely helpful in the management of severe laryngeal papillomatosis. In view of the moderate or absent side effects, prolonged therapy would seem to be feasible.

Skin cancers in two patients with Ev were treated with leukocyte interferon by injection into the lesions and by systemic administration (*Blanchet-Bardon* et al. 1981). Small bowenoid tumors responded very well and regressed, but large invading cancers regressed only in part. It is interesting to note that the benign lesions and their viral contents remained unchanged after systemic administration. This may reflect a virus-type-specific reactivity to interferon.

10 Conclusions

There is a remarkable plurality of human and animal papillomaviruses. The number of virus types is steadily increasing at the moment, although less than 50% cross-hybridization has been defined as the criterion for a new type, which represents one of the most stringent type criteria in virology. A number of observations which seemed to be inconsistent now turn out to reflect properties of different virus types. First of all, specific disease entities correlate with individual HPV types. Additional parameters such as efficiency of transmission, infectivity in the case of impaired host immunity, or accessibility to treatment apparently also correlate with the type of HPV.

In spite of the striking heterogeneity, papillomaviruses reveal a great number of important similarities. They carry group-specific antigens and their DNAs cross-hybridize on the basis of 30% mismatch both in the transforming and in the capsid-coding region. This permits a screening of tumors without knowing which HPV type might be involved.

DNA sequencing of three types disclosed very similar genome structures. There is only one coding strand and all major open reading frames for proteins are of comparable location. Papillomavirus DNA persists extrachromosomally both in transformed cell cultures and in the tumors. Integration into the host genome may happen but does not seem to be essential for transformation. BPV 1-induced transformation apparently depends on virus-specific transcription, which points to a role of viral proteins. However, transforming proteins have not been identified with BPV 1 nor with other papillomavirus types up to now.

Molecular biology has made three major contributions concerning the role of papillomaviruses in malignant conversion: (1) the demonstration of viral DNA in carcinomas of Ev patients, in cervical carcinomas, and in carcinomas of CRPV-infected rabbits; (2) the recognition that only certain types of HPV persist in the carcinomas; and (3) the recognition that malignant conversion in Ev patients correlates with infection by certain HPV types. The last point provides a prospective view and argues against the possibility that the prevalence of specific types merely reflects a preferential pick-up by carcinoma cells. It would be worthwhile following up these three lines of evidence to substantiate the correlation between defined virus types and malignancies. Moreover, a systematic screening of additional squamous cell carcinomas may be likely to detect further neoplasms which reveal papillomavirus fingerprints. This approach would not clarify the role of the viruses in carcinogenesis; indeed, experiments to this end are hard to design. However, the information obtained would certainly be of considerable prognostic value in practice.

Acknowledgments. I am indepted to Drs. E. Amtmann, M.S. Campo, P.M. Howley, G. Orth, E. Schwarz, and H. Zur Hausen, who provided preprints and information on unpublished work. I thank Dr. P. Fuchs for his critical reading of this manuscript, Ingrid Altjohann and Andrea Weiss for their excellent secretarial assistance, and W. Rössler for his skillful photographic work.

Original work cited in this article was supported by the Deutsche Forschungsgemeinschaft.

References

Aaronson DM, Lutzner MA (1967) Epidermodysplasia verruciformis and epidermoid carcinoma. J Am Med Wom Assoc 201:775−777

Almeida JD, Goffe AP (1965) Antibody to wart virus in human sera demonstrated by electron microscopy and precipitin tests. Lancet II:1205−1207

Almeida JD, Howatson AF, Williams MG (1962) Electron microscope study of human warts: Sites of virus production and nature of the inclusion bodies. J Invest Dermatol 38:337−345

Amtmann E, Sauer G (1982a) Activation of non-expressed bovine papilloma virus genomes by tumour promoters. Nature 296:675−677

Amtmann E, Sauer G (1982b) Bovine papilloma virus transcription: Polyadenylated RNA species and assessment of the direction of transcription. J Virol 43:59−66

Amtmann E, Müller H, Sauer G (1980) Equine connective tissue tumors contain unintegrated bovine papilloma virus DNA. J Virol 35:962−964

Archard HO, Heck JW, Stanley HR (1965) Focal epithelial hyperplasia: An unusual oral mucosal lesion found in Indian children. Oral Surg 20:201−212

Barrera-Oro JG, Smith KO, Melnick JL (1962) Quantitation of papovavirus particles in human warts. J Natl Cancer Inst 29:583−595

Barthold SW, Olson C (1978) Common membrane neoantigen on bovine papilloma virus-induced fibroma cells from cattle and horses. Am J Vet Res 39:1643−1645

Benoist C, Chambon P (1981) In vivo sequence requirements of the SV40 early promoter region. Nature 290:304−310

Berman A, Winkelmann RK (1977) Flat warts undergoing involution: Histopathological findings. Arch Dermatol 113:1219−1221

Berman A, Winkelmann RK (1980) Involuting common warts. J Am Acad Dermatol 3:356−362

Berman A, Domnitz JM, Winkelmann RK (1982) Plantar warts recently turned black. Arch Dermatol 118:47−51

Binetruy B, Meneguzzi G, Breathnach R, Cuzin F (1982) Recombinant DNA molecules comprising bovine papilloma virus type 1 DNA linked to plasmid DNA are maintained in a plasmidial state both in rodent fibroblasts and in bacterial cells. EMBO J 1:621−628

Black PH, Hartley JW, Rowe WP, Huebner RJ (1963) Transformation of bovine tissue culture cells by bovine papilloma virus. Nature (Lond) 199:1016−1018

Blanchet-Bardon C, Puissant A, Lutzner M, Orth G, Nutini MT, Guesry P (1981) Interferon treatment of skin cancer in patients with epidermodysplasia verruciformis. Lancet I:274

Boiron M, Levy JP, Thomas M, Friedman JC, Bernard J (1964) Some properties of bovine papilloma virus. Nature (Lond) 201:423−424

Boiron M, Thomas M, Chenialle PH (1965) A biological property of deoxyribonucleic acid extracted from bovine papilloma virus. Virology 52:150−153

Boon ME, Fox CH (1981) Simultaneous condyloma acuminata and dysplasia of the uterine cervix. Acta Cytol 25:393−399

Bosse K, Christophers E (1964) Beitrag zur Epidemiologie der Warzen. Hautarzt 15:80−86

Boyle WF, Riggs JL, Oshiro JS, Lennette EH (1973) Electron microscopic identification of papovavirus in laryngeal papilloma. Laryngoscope 83:1102−1108

Breedis C, Berwick L, Anderson TF (1962) Fractionation of Shope papilloma virus in cesium chloride density gradients. Virology 17:84−94

Breitburd F, Favre M, Zoorob R, Fortin D, Orth G (1981) Detection and characterization of viral genomes and search for tumoral antigens in two hamster cell lines derived from tumors induced by bovine papillomavirus type 1. Int J Cancer 27:693−702

Brodersen I, Genner J (1973) Histological and immunological observations on common warts in regression. Acta Derm Venereol (Stockh) 53:461–464

Bryan WR, Beard JW (1940) Correlation of frequency of positive inoculations with inoculation period and concentration of purified papilloma protein. J Infect Dis 66: 245–253

Bunney MH (1982) Viral warts: Their biology and treatment. Oxford University Press, Oxford

Buschke A, Löwenstein L (1931) Über carcinomähnliche Condylomata acuminata des Penis. Arch Dermatol Syph (Berlin) 163:30–46

Butel JS (1972) Studies with human papilloma virus modeled after known papovavirus systems. J Natl Cancer Inst 48:285–299

Campo MS, Coggins LW (1982) Molecular cloning of bovine papillomavirus genomes and comparison of their sequence homologies by heteroduplex mapping. J Gen Virol 63:255–264

Campo MS, Spandidos DA (1983) Molecularly cloned bovine papillomavirus DNA transforms mouse fibroblasts in vitro. J Gen Virol 64:549–558

Campo MS, Moar MH, Jarrett WFH, Laird HM (1980) A new Papillomavirus associated with alimentary tract cancer in cattle. Nature (Lond) 286:180–182

Campo MS, Moar MH, Laird HM, Jarrett WFH (1981) Molecular heterogeneity and lesion site specificity of cutaneous bovine papillomaviruses. Virology 113:323–335

Campo MS, Spandidos DA, Lang J, Wilkie NM (1983) Transcriptional control signals in the genome of bovine papillomavirus type 1. Nature (Lond) 303:77–81

Chen EY, Howley PM, Levinson AD, Seeburg PH (1982) The primary structure and genetic organization of the bovine papillomavirus type 1 genome. Nature (Lond) 299:529–534

Cheville NF (1966) Studies on connective tissue tumors in the hamster produced by bovine papilloma virus. Cancer Res 26:2334–2339

Chretien JH, Esswein JG, Garagusi VF (1978) Decreased T cell levels in patients with warts. Arch Dermatol 114:213–215

Ciuffo G (1907) Innesto positivo con filtrato di verrucae volgare. Giorn Ital Mal Venereol 48:12–17

Clad A, Gissmann L, Meier B, Freese UK, Schwarz E (1982) Molecular cloning and partial nucleotide sequence of human papillomavirus type 1a DNA. Virology 118: 254–259

Coggin JR Jr, Zur Hausen H (1979) Workshop on papillomaviruses and cancer. Cancer Res 39:545–546

Cohen JC (1980) Methylation of milk-borne and genetically transmitted mouse mammary tumor virus proviral DNA. Cell 19:653–663

Cook T, Cohn A, Brunschwig P, Goepfert H, Butel J, Rawls W (1973) Laryngeal papilloma: Etiologic and therapeutic considerations. Ann Otol Rhinol Laryngeol 82:649–655

Costa J, Howley PM, Bowling MC, Howard R, Bauer WC (1981) Presence of human papilloma viral antigens in juvenile multiple laryngeal papilloma. Am J Clin Pathol 75:194–197

Crawford LV, Crawford EM (1963) A comparative study of polyoma and papilloma viruses. Virology 21:258–263

Croissant O, Testaniere V, Orth G (1982) Mise en évidence et localisation de régions conservées dans les génomes du papillomavirus humain 1a et du papillomavirus bovin 1 par analyse d'"hétéroduplex" au microscope électronique. CR Seances Acad Sci III 294:581–586

Cubie HA (1972) Serological studies in a student population prone to infection with human papilloma virus. J Hyg 70:677–690

Cubie HA (1976) Failure to produce warts on human skin grafts on "nude" mice. Br J Dermatol 94:659–665

Danos O, Katinka M, Yaniv M (1980) Molecular cloning, refined physical map and heterogeneity of methylation sites of papilloma virus type 1a DNA. Eur J Biochem 109:457–461

Danos O, Katinka M, Yaniv M (1982) Human papillomavirus 1a complete DNA sequence: A novel type of genome organization among papovaviridae. EMBO J 1:231–236

Della Torre G, Pilotti S, De Palo G, Rilke F (1978) Viral particles in cervical condylomatous lesions. Tumori 64:549–553

De Peuter M, De Clercq B, Minette A (1977) An epidemiological survey of virus warts of the hands among butchers. Br J Dermatol 96:427–431

Desrosiers RC, Mulder C, Fleckenstein B (1979) Methylation of *Herpes-virus saimiri* DNA in lymphoid tumor cell lines. Proc Natl Acad Sci USA 76:3839–3843

De Villiers EM, Gissmann L, Zur Hausen H (1981) Molecular cloning of viral DNA from human genital warts. J Virol 40:932–935

DiMaio D, Treisman R, Maniatis T (1982) Bovine papillomavirus vector that propagates as a plasmid in both mouse and bacterial cells. Proc Natl Acad Sci USA 79:4030–4034

Dipaolo JA, Popescu NC (1979) Cytogenetics of Syrian hamster and its relationship to in vitro neoplastic transformation. Proc Exp Tumor Res 24:2–16

Dürst M, Gissmann L, Ikenberg H, Zur Hausen H (1983) A new type of papillomavirus DNA from a cervical carcinoma and its prevalence in cancer biopsies from different geographic regions. Proc Natl Acad Sci USA 80:3812–1815

Dvoretzky I, Shober R, Chattopadhy SK, Lowy DR (1980) A quantitative in vitro focus forming assay for bovine papilloma virus. Virology 103:369–375

Evans CA, Ito Y (1966) Antitumor immunity in the Shope papilloma-carcinoma complex of rabbits. III. Response to reinfection with viral nucleic acid. J Natl Cancer Inst 36:1161–1166

Evans CA, Gorman LR, Ito Y, Weiser RS (1962) Antitumor immunity in the SHOPE papilloma-carcinoma complex of rabbits. I. Papilloma regression induced by homologous and autologous tissue vaccines. J Natl Cancer Inst 29:277–285

Fatemi-Nainie S, Anderson LW, Cheevers WP (1982) Identification of a transforming retrovirus from cultured equine dermal fibrosarcoma. Virology 120:490–494

Favre M, Breitburd F, Croissant O, Orth G (1974) Hemagglutinating activity of bovine papilloma virus. Virology 60:572–578

Favre M, Breitburd F, Croissant O, Orth G (1975a) Structural polypeptides of rabbit, bovine, and human papilloma viruses. J Virol 15:1239–1247

Favre M, Orth G, Croissant O, Yanif M (1975b) Human papillomavirus DNA: physical map. Proc Natl Acad Sci USA 72:4810–4814

Favre M, Breitburd F, Croissant O, Orth G (1977) Chromatin-like structures obtained after alkaline disruption of bovine and human papillomaviruses. J Virol 21:1205–1209

Favre M, Jibard N, Orth G (1982) Restriction mapping and physical characterization of the cottontail rabbit papillomavirus genome in transplantable VX2 and VX7 domestic rabbit carcinomas. Virology 119:298–309

Feldman YM, Nikital JA (1979) Condylomata acuminata. NY State J Med 1747–1749

Finch JI, Klug A (1965) The structure of viruses of the papilloma-polyoma type. III. Structure of rabbit papilloma virus. J Mol Biol 13:1–12

Ford JN, Jennings PA, Spradbrow PB, Francis J (1982) Evidence for papillomaviruses in ocular lesions in cattle. Res Vet Sci 32:257–259

Freed DLJ, Eyres KE (1979) Persistent warts protected from immune attack by a blocking factor Br J Dermatol 100:731–733

Freese UK, Schulte P, Pfister H (1982) Papilloma virus-induced tumors contain a virus-specific transcript. Virology 117:257–261

Friedewald WF, Kidd JG (1944) The recoverability of virus from papillomas produced therewith in domestic rabbits. J Exp Med 79:591–605

Friedmann JM, Fialkow PJ (1976) Viral "tumorigenesis" in man: Cell markers in condylomata acuminata. Int J Cancer 17:57–61

Friedman JC, Lewy JP, Lasneret J, Thomas M, Boiron M, Bernard J (1963) Induction de fibromes sous-cutanés chez le hamster doré par inoculation d'extraits à cellulaires de papillomes bovins. CR Seances Acad Sci III 257:2328

Frithiof L, Wersäll J (1967) Virus-like particles in papillomas of the human oral cavity. Arch Virusforschung 21:31−44

Gamperl R, Amtmann E, Pfister H (1983) Chromosome abnormalities in bovine papillomavirus induced hamster tumors and transformed hamster cells. Cancer Res Clin Oncol 105:A37

Genner J (1971) Verruca vulgaris. II. Demonstration of a complement fixation reaction. Acta Derm Venereol (Stockh) 51:365−373

Geraldes A (1969) Malignant transformation of hamster cells by cell-free extracts of bovine papillomas (in vitro). Nature 222:1283−1284

Geraldes A (1970) New antigens in hamster embryo cells transformed in vitro by bovine papilloma extracts. Nature 226:81−82

Gissmann L, Zur Hausen H (1976) Human papilloma viruses: Physical mapping and genetic heterogeneity. Proc Natl Acad Sci USA 73:1310−1313

Gissmann L, Zur Hausen H (1977) Inverted repetitive sequences in human papilloma virus 1 (HPV 1) DNA. Virology 83:271−276

Gissmann L, Zur Hausen H (1980) Partial characterization of viral DNA from human genital warts (condylomata acuminata). Int J Cancer 25:605−609

Gissmann L, Pfister H, Zur Hausen H (1977) Human papilloma viruses (HPV): Characterization of 4 different isolates. Virology 76:569−580

Gissmann L, De Villiers EM, Zur Hausen H (1982a) Analysis of human genital warts (condylomata acuminata) and other genital tumors for human papilloma virus type 6 DNA. Int J Cancer 29:143−146

Gissmann L, Diehl V, Schultz-Coulon HJ, Zur Hausen H (1982b) Molecular cloning and characterization of human papilloma virus DNA derived from a laryngeal papilloma. J Virol 44:393−400

Gissmann L, Wolnik L, Ikenberg H, Koldovsky U. Schnürch HG, Zur Hausen H (1983) Human papillomavirus types 6 and 11 DNA sequences in genital and laryngeal papillomas and in some cervical cancers. Proc Natl Acad Sci USA 80:560−563

Glinski W, Jablonska S, Langner A, Obalek S, Haftek M, Proniewska M (1976) Cell-mediated immunity in epidermodysplasia verruciformis. Dermatologica 153:218−227

Glinski W, Obalek S, Jablonska S, Orth G (1981) T cell defect in patients with epidermodysplasia verruciformis due to human papillomavirus type 3 and 5. Dermatologica 162:141−147

Göbel U, Arnold W, Wahn VM, Treuner J, Jürgens H, Cantell K (1981) Comparison of human fibroblast and leukocyte interferon in the treatment of severe laryngeal papillomatosis in children. Eur J Pediatr 137:175−176

Goepfert H, Gutterman JU, Dichtel WJ, Sessions RB, Cangir A, Sulek M (1982) Leukocyte interferon in patients with juvenile laryngeal papillomatosis. Ann Otol Rhinol Laryngol 91:431−436

Green M, Orth G, Wold WSM, Sanders PR, Mackey JK, Favre M, Croissant O (1981) Analysis of human cancers, normal tissues, and verrucae plantares for DNA sequences of human papillomavirus types 1 and 2. Virology 110:176−184

Green M, Brackmann KH, Sanders PR, Loewenstein PM, Freel JH, Eisinger M, Switlyk SA (1982) Isolation of a human papillomavirus from a patient with epidermodysplasia verruciformis: Presence of related viral DNA genomes in human urogenital tumors. Proc Natl Acad Sci USA 79:4437−4441

Greenberg MJH, Smith MTL, Katz RM (1973) Verrucae vulgaris rejection. A preliminary study of contact dermatitis and cellular immunity response. Arch Dermatol 107:580−582

Gross G, Pfister H, Gissmann L, Hagedorn M (1982) Correlation between human papillomavirus type (HPV) and histology of warts. J Invest Dermatol 78:160−164

Gross G, Pfister H, Hagedorn M, Stahn R (1983) Effect of oral aromatic retinoid (Ro 10-9359) on human papillomavirus 2 induced common warts. Dermatologica 166:48–53

Gross GE, Pfister H, Mittermayer C (1980) Papilloma virus particles in a fibroma of the tongue. Acta Derm Venereol (Stockh) 60:315–318

Gruss P, Dhar R, Khoury G (1981) Simian virus 40 tandem repeated sequences as an element of the early promoter. Proc Natl Acad Sci USA 78:943–947

Grussendorf EI (1980) Lichtmikroskopische Untersuchungen an typisierten Virus-warzen (HPV-1 und HPV-4). Arch Dermatol Res 268:141–148

Grussendorf EI, Gahlen W (1975) Metaplasia of a verruca vulgaris into spinocellular carcinoma. Dermatologica 150:295–299

Grussendorf EI, Zur Hausen H (1979) Localization of viral DNA-replication in sections of human warts by nucleic acid hybridization with complementary RNA of human papilloma virus type 1. Arch Dermatol Res 264:55–63

Haglund S, Lundquist PG, Cantell K, Strander H (1981) Interferon therapy in juvenile laryngeal papillomatosis. Arch Otolaryngol 107:327–332

Hayward WS, Neel BG, Astrin SM (1981) Activationof cellular oncogene by promoter insertion in ALV-induced lymphoid leukosis. Nature 290:475–480

Heilman CA, Law MF, Israel MA, Howley PM (1980) Cloning of human papilloma virus genomic DNAs and analysis of homologous polynucleotide sequences. J Virol 36:395–407

Heilman CA, Engel L, Lowy DR, Howley PM (1982) Virus-specific transcription in bovine papillomavirus transformed mouse cells. Virology 119:22–34

Hellström I, Evans CA, Hellström KE (1969) Cellular immunity and its serum-mediated inhibition in Shope-virus-induced rabbit papillomatosis. Int J Cancer 4:601–607

Hills E, Laverty CR (1979) Electron microscopic detection of papilloma virus particles in selected koilocytotic cells in a routine cervical smear. Acta Cytol 23:53–56

Howley PM, Law MF, Heilman C, Engel L, Alonso MC, Israel MA, Lowy DR, Lancaster WD (1980) Molecular characterization of papillomavirus genomes. Cold Spring Harbor Conf Cell Prolif 7:233–247

Hoxtell EO, Mandel JS, Murray SS, Schuman LM, Goltz RW (1977) Incidence of skin carcinoma after renal transplantation. Arch Dermatol 113:436–438

Ishimoto A, Ito Y (1969) Specific surface antigens in Shope papilloma cells. Virology 39:595–597

Ishimoto A, Oota S, Kimura I, Miyake T, Ito Y (1970) In vitro cultivation and anti-genicity of cottontail rabbit papilloma cells induced by the Shope papilloma virus. Cancer Res 30:2598–2605

Ito Y (1963) Studies on subviral tumorigenesis: carcinoma derived from nucleic acid-induced papillomas of rabbit skin. Acta Unio Int Contra Cancrum 19:280–283

Ito Y (1975) Papilloma-myxoma viruses. In: Becker FF (ed) Cancer, vol 2. Plenum Press, New York, pp 323–341

Ito Y (1981) Effect of an aromatic retinoic acid analog (Ro 10–9359) on growth of virus-induced papilloma (Shope) and related neoplasia of rabbits. Eur J Cancer 17: 35–42

Jablonska S, Biczysko W, Jakubowicz W, Dabrowski H (1970) The ultrastructure of transitional states to Bowen's disease and invasive Bowen's carcinoma in epidermo-dysplasia verruciformis. Dermatologica 140:186–194

Jablonska S, Dabrowski J, Jakubowicz K (1972) Epidermodysplasia verruciformis as a model in studies on the role of papova viruses in oncogenesis. Cancer Res 32: 583–589

Jablonska S, Orth G, Jarzabek-Chorzelska M, Rzesa G, Obalek S, Glinski W, Favre M, Croissant O (1979) Epidermodysplasia verruciformis versus disseminated verrucae planae: Is epidermodysplasia verruciformis a generalized infection with wart virus? J Invest Dermatol 72:114–119

Jablonska S, Orth G, Glinski G, Obalek S, Jarzabek-Chorzelska M, Croissant O, Favre M, Rzesa G (1980) Morphology and immunology of human warts and familial warts. In: Bachmann PA (ed) Leukaemias, lymphomas and papillomas: Comparative aspects. Taylor and Francis, London, pp 107–131

Jablonska S, Obalek S, Wolska H, Jarzabek-Chorzelska M (1981) Ro 10-9359 in epidermodysplasia verruciformis. In: Orfanos CE, Braun-Falco O, Farber EM, Grupper Ch, Polano MK, Schuppli R (eds) Retinoids, advances in basic research and therapy. Springer, Berlin Heidelberg New York, pp 401–405

Jablonska S, Obalek S, Orth G, Haftek M, Jarzabek-Chorzelska M (1982) Regression of the lesions of epidermodysplasia verruciformis. Br J Dermatol 107:109–116

Jacyk WK, Subbuswamy SG (1979) Epidermodysplasia verruciformis in Nigerians. Dermatologica 159:256–265

Jarrett WFH (1980) Bovine papillomaviruses and alimentary malignancy. In: Bachmann PA (ed) Leukaemia, lymphomas and papillomas: Comparative aspects. Taylor and Francis, London, pp 87–91

Jarrett WFH, McNeil PE, Grimshaw WTR, Selman IE, McIntyre WIM (1978) High incidence area of cattle cancer with a possible interaction between an environmental carcinogen and a papilloma virus. Nature 274:215–217

Jenson AB, Rosenthal JR, Olson C, Pass F, Lancaster WD, Shah K (1980) Immunological relatedness of papillomaviruses from different species. J Natl Cancer Inst 64: 495–500

Kaufmann J, Meves C, Ott F (1978) Die Epidermodysplasia verruciformis Lewandowsky-Lutz in licht- und elektronenoptischem Vergleich mit den übrigen Papova-Virus-Akanthomen. Arch Dermatol Res 261:39–54

Kidd JG (1938) The course of virus-induced rabbit papillomas as determined by virus cells and host. J Exp Med 67:551–574

Kidd JG, Rous P (1940) Cancers deriving from the virus papillomas of wild rabbits under natural conditions. J Exp Med 71:469–485

Kidd JG, Beard JW, Rous P (1936) Serological reactions with a virus causing a rabbit papilloma which becomes cancerous. II. Tests of the blood of animal carrying various tumors. J Exp Med 64:63–78

Kienzler J-L, Laurent R, Coppey J, Favre M, Orth G, Coupez L, Agache P (1979) Epidermodysplasie verruciforme. Données ultrastructurales, virologiques et photobiologiques; à propos d'une observation. Ann Dermatol Venereol 106:549–563

Kimura S, Hirai A, Harada R, Nagashima M (1978) So-called multicentric pigmented Bowen's disease. Dermatologica 157:229–237

Kleinsasser O, Oliveira e Cruz G (1973) "Juvenile" und "adulte" Kehlkopfpapillome. HNO 21:97–106

Kleinschmidt AK, Kass SJ, Williams RC, Knight CA (1965) Cyclic DNA of Shope papilloma virus. J Mol Biol 13:749–756

Klug A, Finch JT (1965) Structure of virus of the papilloma-polyoma type I. Human wart virus. J Mol Biol 11:403–423

Koller LD, Olson C (1972) Attempted transmission of warts from man, cattle, and horses, and of deer fibroma, to selected hosts. J Invest Dermatol 58:366–368

Koranda FC, Dehmel EM, Kahn G, Penn I (1974) Cutaneous complications in immunosuppressed renal homograft recipients. JAMA 229:419–424

Kreider JW (1963) Studies on the mechanism responsible for the spontaneous regression of the Shope rabbit papilloma. Cancer Res 23:1593–1599

Kreider JW (1980) Neoplastic progression of the Shope rabbit papilloma. Cold Spring Harbor Conf Cell Prolif 7:283–300

Kremsdorf D, Jablonska S, Favre M, Orth G (1982) Biochemical characterization of two types of human papillomaviruses associated with epidermodysplasia verruciformis. J Virol 43:436–447

Krzyzek RA, Watts SL, Anderson DL, Faras AJ, Pass F (1980) Anogenital warts contain several distinct species of human papillomavirus. J Virol 36:236–244

Kurman RJ, Shah KH, Lancaster WD, Jenson AB (1981) Immunoperoxidase localization of papillomavirus antigens in cervical dysplasia and vulvar condylomas. Am J Obstet Gynecol 140:931–935

Kurman RJ, Sanz LE, Jenson AB, Perry S, Lancaster WD (1982) Papillomavirus infection of the cervix I. Correlation of histology with viral structural antigens and DNA sequences. Int J Gynecol Pathol 1:17–28

Lack EE, Jenson AB, Smith HG, Healy GB, Pass F, Vawter GF (1980) Immunoperoxidase localization of human papillomavirus in laryngeal papillomas. Intervirology 14:148–154

Lancaster WD (1979) Physical maps of bovine papillomavirus type 1 and type 2 genomes. J Virol 32:684–687

Lancaster WD (1981) Apparent lack of integration of bovine papillomavirus DNA in virus-induced equine and bovine tumors and virus-transformed mouse cells. Virology 108:251–255

Lancaster WD, Jenson AB (1981) Evidence for papillomavirus antigens and DNA sequences in laryngeal papilloma. Intervirology 15:204–212

Lancaster WD, Meinke W (1975) Persistence of viral DNA in human cell cultures infected with human papilloma virus. Nature 256:434–436

Lancaster WD, Olson C (1978) Demonstration of two distinct classes of bovine papilloma virus. Virology 89:371–379

Lancaster WD, Olson C (1980) State of bovine papilloma virus DNA in connective tissue tumors. Cold Spring Harbor Conf Cell Prolif 7:223–232

Lancaster WD, Olson C (1982) Animal papillomaviruses. Microbiol Rev 46:191–207

Lancaster WD, Sundberg JP (1982) Characterization of papilloma viruses isolated from cutaneous fibromas of white-tailed deer and mule deer. Virology 123:212–216

Lancaster WD, Olson C, Meinke W (1976) Quantitation of bovine papilloma viral DNA in viral-induced tumors. J Virol 17:824–831

Lancaster WD, Olson C, Meinke W (1977) Bovine papilloma virus: presence of virus-specific DNA sequences in naturally occurring equine tumors. Proc Natl Acad Sci USA 74:524–528

Lancaster WD, Theilen GH, Olson C (1979) Hybridization of bovine papilloma virus type 1 and type 2 DNA to DNA from virus-induced hamster tumors and naturally occurring equine connective tissue tumors. Intervirology 11:227–233

LaPorta RF, Taichman LB (1982) Human papilloma viral DNA replicates as a stable episome in cultured epidermal keratinocytes. Proc Natl Acad Sci USA 79:3393–3397

Laurent R, Coume-Marquet S, Kienzler JL, Lambert D, Agache P (1978) Comparative electron microscopic study of clear cells in epidermodysplasia verruciformis and flat warts. Arch Dermatol Res 263:1–12

Laverty CR, Russell P, Hills E, Booth N (1978) The significance of noncondylomatous wart virus infection of the cervical transformation zone. A review with discussion of two illustrative cases. Acta Cytol 22:195–201

Law M-F, Lancaster WD, Howley PM (1979) Conserved sequences among the genomes of papillomaviruses. J Virol 32:199–207

Law M-F, Lowy DR, Dvoretzky I, Howley PM (1981) Mouse cells transformed by bovine papillomavirus contain only extrachromosomal viral DNA sequences. Proc Natl Acad Sci USA 78:2727–2731

Law M-F, Howard B, Sarver N, Howley PM (1982) Expression of selective traits in mouse cells, transformed with a BPV DNA derived hybrid molecule containing E. coli gpt. In: Gluzman Y (ed) Eukaryotic viral vectors. Cold Spring Harbor Laboratory, Cold Spring Harbor, New York, pp 79–85

Le Bouvier GL, Sussman M, Crawford LV (1966) Antigenic diversity of mammalian papillomaviruses. J Gen Microbiol 45:497–501

Lecatsas G, Boes E (1979) Papillomavirus in pregnancy urine. Lancet II:533–534

Lee AKY, Eisinger M (1976) Cell-mediated immunity (CMI) to human wart virus and wart-associated tissue antigens. Clin Exp Immunol 26:419–424

Lewandowsky F, Lutz W (1922) Ein Fall einer bisher nicht beschriebenen Hauterkrankung (Epidermodysplasia verruciformis). Arch Dermatol Syph (Berlin) 141:193–203

Lowy DR, Dvoretzky I, Shober R, Law M-F, Engel L, Howley PM (1980) In vitro tumorigenic transformation by a defined subgenomic fragment of bovine papilloma virus DNA. Nature (Lond) 287:72–74

Lusky M, Botchan M (1981) Inhibition of SV40 replication in simian cells by specific pBR322 DNA sequences. Nature 293:79–81

Lusky M, Berg L, Botchan M (1982) Enhancement of tk transformation by sequences of bovine papilloma virus. In: Gluzman Y (ed) Eukaryotic viral vectors. Cold Spring Harbor Laboratory, Cold Spring Harbor, New York

Lutzner MA (1978) Epidermodysplasia verruciformis. Bull Cancer 65:169–182

Lutzner MA, Blanchet-Bardon C (1980) Oral retinoid treatment of human papillomavirus type 5-induced epidermodysplasia verruciformis. N Engl J Med 302:1091

Lutzner M, Croissant O, Ducasse M-F, Kreis H, Crosnier J, Orth G (1980) A potentially oncogenic human papillomavirus (HPV-5) found in two renal allograft recipients. J Invest Dermatol 75:353–356

Massing AM, Epstein WL (1963) Natural history of warts. A two-year study. Arch Dermatol 87:306

Matthews REF (1982) Classification and nomenclature of viruses. Intervirology 17:1–199

Matthews RS, Shirodaria PV (1973) Study of regressing warts by immunofluorescence. Lancet I:689–691

Mayer H, Bollag W, Hänni R, Rüegg R (1978) Retinoids, a new class of compounds with prophylactic and therapeutic activities in oncology and dermatology. Experientia 34:1105–1119

McMichael H (1965) Inhibition of growth of Shope papilloma by hypervitaminosis A. Cancer Res 25:947–956

McMichael H (1967) Inhibition by methylprednisolone of regression of the Shope rabbit papilloma. J Natl Cancer Inst 39:55–65

McVay P, Fretz M, Wettstein F, Stevens J, Ito Y (1982) Integrated Shope virus DNA is present and transcribed in the transplantable rabbit tumour Vx-7. J Gen Virol 60:271–278

Meinke W, Meinke GC (1981) Isolation and characterization of the major capsid protein of bovine papilloma virus type 1. J Gen Virol 52:15–24

Meischke HRC (1979) In vitro transformation by bovine papilloma virus. J Gen Virol 43:471–487

Meisels A, Morin C, Casas-Cordero M (1982) Human papillomavirus infection of the uterine cervix. Int J Gynecol Pathol 1:75–94

Moar MH, Campo MS, Laird HM, Jarrett WFH (1981a) Unintegrated viral DNA sequences in hamster tumor induced by bovine papilloma virus. J Virol 39:945–949

Moar MH, Campo MS, Laird H, Jarrett WFH (1981b) Persistence of nonintegrated viral DNA in bovine cells transformed in vitro by bovine papillomavirus type 2. Nature 293:749–751

Moreno-Lopez J, Pettersson U, Dinter Z, Philipson L (1981) Characterization of a papilloma virus from the European elk (EEPV). Virology 112:589–595

Morgan DM, Meinke W (1980) Isolation of clones of hamster embryo cells transformed by the bovine papilloma virus. Current Microbiology 3:247–251

Morin C, Meisels A (1980) Human papilloma virus, infectin of the uterine cervix. Acta Cytol 24:82–84

Morin C, Braun L, Casas-Cordero M, Shah KV, Roy M, Fortier M, Meisels A (1981) Confirmation of the papillomavirus etiology of condylomatous cervix lesions by the peroxidase-antiperoxidase technique. J Natl Cancer Inst 66:831–834

Morison WL (1975a) Viral warts, herpes simplex and herpes zoster in patients with secondary immune deficiencies and neoplasms. Br J Dermatol 92:625–630

Morison WL (1975b) Cell-mediated immune response in patients with warts. Br J Dermatol 93:553–556

Morison WL (1975c) In vitro assay of immunity to human wart antigen. Br J Dermatol 93:545–551

Mounts P, Shah KV, Kashima H (1982) Viral etiology of juvenile- and adult-onset squamous papilloma of the larynx. Proc Natl Acad Sci USA 79:5425–5429

Müller H, Gissmann L (1978) *Mastomys natalensis* papilloma virus (MnPV), the causative agent of epithelial proliferations: Characterization of the virus particle. J Gen Virol 41:315–323

Mullen DL, Silverberg SG Penn I, Hammond WS (1976) Squamous cell carcinoma of the skin and lip in renal homograft recipients. Cancer 37:729–734

Murphy MF, Potter HL, Abraham JM, Morgan DM, Meinke WJ (1981) Analysis of a restriction endonuclease map for a rabbit papillomavirus DNA. Current Microbiology 5:349–352

Murray RF, Hobbs J, Payne B (1971) Possible clonal origin of common warts (verruca vulgaris). Nature (Lond) 232:51–52

Nasseri M, Wettstein FO, Stevens JG (1982) Two colinear and spliced viral transcripts are present in non-virus-producing benign and malignant neoplasms induced by the Shope (rabbit) papilloma virus. J Virol 44:263–268

Noyes WF (1965) Structure of the human wart virus. Virology 23:65–72

Nürnberger F, Pfister H, Zur Hausen H (1981) Epidermodysplasia verruciformis bei einem Westafrikaner, verursacht durch einen bisher unbekannten Typ eines humanen Papillomavirus (HPV 8) – Versuch einer oralen Retinoidbehandlung. Hautarzt 32 (Suppl V):480

Obalek S, Glinski W, Haftek M, Orth G, Jablonska S (1980) Comparative studies on cell-mediated immunity in patients with different warts. Dermatologica 161:73–83

Ogilvie MM (1970) Serological studies with human papova (wart) virus. J Hyg 68: 479–483

Oguchi M, Komura J, Tagami H, Ofuji S (1981a) Ultrastructural studies of spontaneously regressing plane warts. Langerhans cells show marked activation. Arch Dermatol Res 271:55–61

Oguchi M, Komura J, Tagami H, Ofuji S (1981b) Ultrastructural studies of spontaneously regressing plane warts. Macrophages attack verruca-epidermal cells. Arch Dermatol Res 270:403–411

Olson C, Cook RH (1951) Cutaneous sarcoma-like lesions of the horse caused by the agent of bovine papilloma. Proc Soc Exp Biol Med 77:281–284

Olson C, Gordon DE, Robl MG, Lee KP (1969) Oncogenicity of bovine papilloma virus. Arch Environ Health 19:827–837

Orfanos CE, Strunk W, Gartmann H (1974) Fokale epitheliale Hyperplasie der Mundschleimhaut: Hecksche Krankheit. Dermatologica 149:163–175

Oriel JD (1971a) Natural history of genital warts. Br J Vener Dis 47:1–8

Oriel JD (1971b) Anal warts and anal coitus. Br J Vener Dis 47:373–376

Orth G, Croissant O (1968) Charactères des cultures de première explantation de cellules de papillomes provoqués par le virus de Shope chez le lapin domestique. CR Seances Acad Sci III 266:1084–1087

Orth G, Jeanteur P, Croissant O (1971) Evidence for localization of vegetative viral DNA replication by autoradiographic detection of RNA-DNA hybrids in sections of tumors induced by Shope papilloma virus. Proc Natl Acad Sci USA 68:1876–1880

Orth G, Favre M, Croissant O (1977) Characterization of a new type of human papilloma virus that causes skin warts. J Virol 24:108–120

Orth G, Breitburd F, Favre M (1978a) Evidence for antigenic determination shared by the structural polypeptides of (Shope) rabbit papillomavirus and human papillomavirus type 1. Virology 91:243–255

Orth G, Jablonska S, Favre M, Croissant O, Jarzabek-Chorzelska M, Rzesa G (1978b) Characterization of two new types of human papilloma viruses in lesions of epidermodysplasia verruciformis. Proc Natl Acad Sci USA 75:1537–1541

Orth G, Jablonska S, Jarzabek-Chorzelska M, Obalek S, Rzesa G, Favre M, Croissant O (1979) Characteristics of the lesions and risk of malignant conversion associated with the type of human papillomavirus involved in epidermodysplasia verruciformis. Cancer Res 39:1074:1082

Orth G, Favre M, Breitburd F, Croissant O, Jablonska S, Obalek S, Jarzabek-Chorzelska M, Rzesa G (1980) Epidermodysplasia verruciformis: A model for the role of papilloma viruses in human cancer. Cold Spring Harbor Conf Cell Prolif 7:259–282

Orth G, Jablonska S, Favre M, Croissant O, Obalek S, Jarzabek-Chorzelska M, Jibard N (1981) Identification of papillomaviruses in butcher's warts. J Invest Dermatol 76:97–102

Osterhaus AMDE, Ellens DJ, Horzinek MC (1977) Identification and characterization of a papillomavirus from birds (Fringillidae). Intervirology 8:351–359

Ostrow RS, Krzyzek R, Pass F, Faras AJ (1981) Identification of a novel human papilloma virus in cutaneous warts of meat handlers. Virology 108:21–27

Ostrow RS, Bender M, Niimura M, Seki T, Kawashima M, Pass F, Faras AJ (1982) Human papillomavirus DNA in cutaneous primary and metastasized squamous cell carcinomas from patients with epidermodysplasia verruciformis. Proc Natl Acad Sci USA 79:1634–1638

Parsons RJ, Kidd JG (1942) Oral papillomatosis of rabbits: A virus disease. J Exp Med 77:233–250

Pass F, Maizel JV (1973) Wart-associated antigens. II. Human immunity to viral structural proteins. J Invest Dermatol 60:307–311

Pass F, Marcus DM (1973) Wart-associated antigens. I. Isolation of tissue antigens using antibody immunoadsorbents. J Invest Dermatol 60:301–306

Pass F, Niimura M, Kreider JW (1973) Prolonged survival of human skin xenografts on antithymocyte serum-treated mice: failure to produce verrucae by inoculation with extracts of human warts. J Invest Dermatol 61:371–374

Pathak S, Hsu TC, Trentin JJ, Butel JS, Panigrahy B (1981) Nonrandom chromosome abnormalities in transformed syrian hamster cell lines. In: Arrighi FE, Rao PN, Stubblefield E (eds) Genes, chromosomes and neoplasia. Raven Press, New York, pp 405–417

Perry TL, Harman L (1974) Warts in diseases with immune defects. Cutis 13:359–362

Petzoldt D, Pfister H (1980) HPV 1 DNA in lesions of focal epithelial hyperplasia Heck. Arch Dermatol Res 268:313–314

Pfister H (1980) Comparative aspects of papillomatosis. In: Bachmann PA (ed) Leukaemias, lymphomas and papillomas: Comparative aspects. Taylor and Francis, London, pp 93–106

Pfister H, Meszaros J (1980) Partial characterization of a canine oral papillomavirus. Virology 104:243–246

Pfister H, Zur Hausen H (1978a) Characterization of proteins of human papilloma viruses (HPV) and antibody response to HPV 1. Med Microbiol Immunol 166:13–19

Pfister H, Zur Hausen H (1978b) Seroepidemiological studies of human papilloma virus (HPV 1) infections. Int J Cancer 21:161–165

Pfister H, Gissmann L, Zur Hausen H (1977) Partial characterization of proteins of human papilloma viruses (HPV) 1–3. Virology 83:131–137

Pfister H, Gross G, Hagedorn M (1979a) Characterization of human papillomavirus 3 in warts of a renal allograft patient. J Invest Dermatol 73:349–353

Pfister H, Huchthausen B, Gross G, Zur Hausen H (1979b) Seroepidemiological studies of bovine papillomavirus infections. J Natl Cancer Inst 62:1423–1425

Pfister H, Linz U, Gissmann L, Huchthausen B, Hoffmann D, Zur Hausen H (1979c) Partial characterization of a new type of bovine papillomaviruses. Virology 96:1–8

Pfister H, Gissmann H, Zur Hausen H, Gross G (1980) Characterization of human and bovine papilloma viruses and of the humoral immune response to papilloma virus infection. Cold Spring Harbor Conf Cell Prolif 7:249—258

Pfister H, Fink B, Thomas C (1981a) Extrachromosomal bovine papillomavirus type 1 DNA in hamster fibromas and fibrosarcomas. Virology 115:414—418

Pfister H, Nürnberger F, Gissmann L, Zur Hausen H (1981b) Characterization of a human papillomavirus from epidermodysplasia verruciformis lesions of a patient from Upper Volta. Int J Cancer 27:645—650

Pfister H, Gassenmaier A, Nürnberger F, Stüttgen G (1983a) HPV 5 DNA in a carcinoma of epidermodysplasia verruciformis patient infected with various human papillomavirus types. Cancer Res 43:1436—1441

Pfister H, Hettich I, Runne U, Gissmann L, Chilf GN (1983b) Characterization of human papillomavirus type 13 from lesions of focal epithelial hyperplasia Heck. J Virol 47:363—366

Prawer SE, Pass F, Vance JC, Greenberg LJ, Yunis EJ, Zelickson AS (1977) Depressed immune functions in epidermodysplasia verruciformis. Arch Dermatol 113:495—499

Puget A, Favre M, Orth G (1975) Induction de tumeurs fibroplastiques cutanées où sous-cutanées chez l'ochotone afghan *(Ochotono rufescens rufescens)* par inoculation du virus du papillome bovin. CR Seances Acad Sci III 280:2813—2816

Pyrhönen S (1978) Human wart antibodies in patients with genital and skin warts. Acta Derm Venereol (Stockh) 58:427—432

Pyrhönen S, Johansson E (1975) Regression of warts. An immunological study. Lancet I:592—596

Pyrhönen S, Penttinen K (1972) Wart virus antibodies and the prognosis of wart disease. Lancet II:1330—1332

Rangan SRS, Gutter A, Baskin GB, Anderson D (1980) Virus associated papillomas in colobus monkeys *(Colobus guereza).* Lab Anim Sci 30(5):885—889

Rattanapanone V, Tashiro S, Tokuda H, Rattanapanone N, Ito Y (1981) Cellular retinoic acid-binding protein in virus-induced Shope papillomas of rabbit skin. Cancer Res 41:1483—1487

Reid TMS, Fraser NG, Kernohan IR (1976) Generalized warts and immune deficiency. Br J Dermatol 95:559—564

Reinacher M, Müller H, Thiel W, Rudolph RL (1978) Localization of papillomavirus and virus-specific antigens in the skin of tumor-bearing *Mastomys natalensis* (GRA Giessen). Med Microbiol Immunol 165:93—99

Rheinwald JG, Green H (1975) Serial cultivation of strains of human epidermal keratinocytes: the formation of keratinizing colonies from single cells. Cell 6:331—344

Robl MG, Olson C (1968) Oncogenic action of bovine papilloma virus in hamsters. Cancer Res 28:1596—1604

Rogers S, Rous P (1951) Joint action of a chemical carcinogen and a neoplastic virus to induce cancer in rabbits. Results of exposing epidermal cells to a carcinogenic hydrocarbon at time of infection with the Shope papilloma virus. J Exp Med 93:459—488

Rogers S, Kidd JG, Rous P (1960) Relationships of the Shope papilloma virus to cancers it determines in domestic rabbits. Acta Unio Int Contra Cancrum 16:129—130

Rotkin ID (1973) A comparison review of key epidemiological studies in cervical cancer related to current searches for transmissible agents. Cancer Res 33:1353—1367

Rous P, Beard JW (1934) A virus-induced mammalian growth with the characters of a tumor (the Shope rabbit papilloma). J Exp Med 60:701—722

Rous P, Beard JW (1935) The progression to carcinoma of virus-induced rabbit papillomas (Shope). J Exp Med 65:523—548

Rous P, Friedewald WF (1944) The effect of chemical carcinogens on virus-induced rabbit papillomas. J Exp Med 79:511—537

Rous P, Kidd JG (1938) The carcinogenic effect of a papilloma virus on the tarred skin of rabbits. I. Description of the phenomenon. J Exp Med 67:399–422

Rowson KEK, Mahy BWJ (1967) Human papova (wart) virus. Bacteriol Rev 31:110–131

Ruiter M, van Mullem PJ (1970) Behaviour of virus in malignant degeneration of skin lesions in epidermodysplasia verruciformis. J Invest Dermatol 54:324–331

Rulison RH (1942) Warts. A statistical study of nine hundred twenty-one cases. Arch Dermatol Syphilol 46:66–81

Sarver N, Gruss P, Law MF, Khoury G, Howley PM (1981) Bovine papilloma virus deoxyribonucleic acid: a novel eucaryotic cloning vector. Mol Cell Biol 1:486–496

Sarver N, Byrne JC, Howley PM (1982) Transformation and replication in mouse cells of a bovine papillomavirus-pML2 plasmid vector that can be rescued in bacteria. Proc Natl Acad Sci USA 79:7147–7151

Schaffhausen BS, Benjamin TL (1976) Deficiency in histone acetylation in nontransforming host range mutants of polyoma virus. Proc Natl Acad Sci USA 73:1092–1096

Schmauz R, Owor R (1980) Epidemiology of malignant degeneration of condylomata acuminata in Uganda. Pathol Res Pract 170:91–103

Schwarz et al. (to be published) Nucleotide sequence of HPV 6 B-DNA.

Seski JC, Reinhalter ER, Silva J Jr (1977) Abnormalities of lymphocyte transformations in women with condylomata acuminata. Obstet Gynecol 51:188–192

Seto A, Notake K, Kawanishi M, Ito Y (1977) Development and regression of Shope papillomas induced in newborn domestic rabbits. Proc Soc Exp Biol Med 156:64–67

Shah KH, Lewis MG, Jenson AB, Kurman RJ, Lancaster WD (1980) Papillomavirus and cervical dysplasia. Lancet II:1190

Shelley WB, Wood MG (1981) Transformation of the common wart into squamous cell carcinoma in a patient with primary lymphedema. Cancer 48:820–824

Shirodaria PV, Matthews RS (1975) An immunofluorescence study of warts. Clin Exp Immunol 21:329–338

Shope RE, Mangold R, McNamara LG, Dumbell KR (1958) An infectious cutaneous fibroma of the Virginia white-tailed deer *(Odocoileus virginianus)*. J Exp Med 108:797–802

Siegel SE, Isaacs H Jr, Cohen SR, Stanley P (1979) Malignant transformation of tracheobronchial juvenile papillomatosis without prior radiotherapy. Ann Otol Rhinol Laryngol 88:192–197

Spencer ES, Andersen HK (1970) Clinically evident non-terminal infections with herpesviruses and the wart virus in immunosuppressed renal allograft patients. Br Med J III:251–254

Spradbrow PB, Hoffmann D (1980) Bovine ocular squamous cell carcinoma. Vet Bull 50:449–459

Spradbrow PB, Beardmore GL, Francis J (1983) Virions resembling papillomaviruses in hyperkeratotic lesions from sun-damaged skin. Lancet I:189

Starzl TE, Porter KA, Andres G, Halgrimson CG, Hurwitz R, Giles G, Terasaki PJ, Penn J, Schroter GT, Lilly J, Starkie SJ, Putnam CW (1970) Long-term survival after renal transplantation in humans. Ann Surg 172:437–472

Steigleder GK (1978) Histology of benign virus induced tumors of the skin. J Cutan Pathol 5:45–52

Stevens JG, Wettstein FO (1979) Multiple copies of Shope virus DNA are present in cells of benign and malignant non-virus-producing neoplasms. J Virol 30:891–898

Strauss MJ, Bunting H, Melnick JL (1950) Virus-like particles and inclusion bodies in skin papillomas. J Invest Dermatol 15:433–443

Sundberg JP, Russell WC, Lancaster W (1981) Papillomatosis in Indian elephants. JAMA 179:1247–1248

Syrjänen KJ (1979) Morphologic survey of the condylomatous lesions in dysplastic and neoplastic epithelium of the uterine cervix. Arch Gynecol 227:153–161

Syrjänen KJ (1980) Bronchial squamous cell carcinomas associated with epithelial changes identical to condylomatous lesions of the uterine cervix. Lung 158:131−142

Syrjänen KJ (1982) Histological changes identical to those of condylomatous lesions found in esophageal squamous cell carcinomas. Arch Geschwulstforsch 52:383−392

Syverton JT (1952) The pathogenesis of the rabbit papilloma-to-carcinoma sequence. Ann NY Acad Sci 54:1126−1140

Tagami H, Takigawa M, Ogino A, Imamura S, Ofugi S (1977) Spontaneous regression of plane warts after inflammation. Arch Dermatol 113:1209−1213

Takigawa M, Tagami H, Watanabe S, Ogino A, Imamura S, Ofugi S (1977) Recovery processes during regression of plane warts. Arch Dermatol 113:1214−1218

Thivolet J, Viac J (1978) Immunologie des verrues humaines. Ann Dermatol Venereol (Paris) 105:257−264

Thomas M, Levy JP, Tanzer J, Boiron M, Bernard J (1963) Transformation in vitro de cellules de peau de veau embryonnaire sous l'action d'extrants a cellulaires de papillomes bovins. CR Seances Acad Sci III 257:2155−2158

Thomas M, Boiron M, Tanzer J, Levy JP, Bernard J (1964) In vitro transformation of mice cells by bovine papilloma virus. Nature (Lond) 202:709−710

Tooze J (1980) DNA tumor viruses. Molecular biology of tumor viruses, part 2. Cold Spring Harbor Laboratory, Cold Spring Harbor, New York

Turek LP, Byrne JC, Lowy DR, Dvoretzky I, Friedman RM, Howley PM (1982) Interferon induces morphologic reversion with elimination of extrachromosomal viral genomes in bovine papillomavirus-transformed mouse cells. Proc Natl Acad Sci USA 79:7914−7918

Vanselow BA, Spradbrow PB (1982) Papillomaviruses, papillomas and squamous cell carcinomas in sheep. Vet Rec 110: 561−562

van Wyk CW (1977) Focal epithelial hyperplasia of the mouth: recently discovered in South Africa, Br J Dermatol 96:381−388

Viac J, Thivolet J, Chardonnet Y (1977a) Specific immunity in patients suffering from recurring warts before and after repetitive intradermal tests with human papilloma virus. Br J Dermatol 97:365−370

Viac J, Thivolet J, Hegazy MR, Chardonnet Y, Dambuyant C (1977b) Comparative study of delayed hypersensitivity skin reactions and antibodies to human papilloma virus (HPV). Clin Exp Immunol 29:240−246

Viac J, Schmitt D, Thivolet J (1978) An immunoelectron microscopic localization of wart associated antigens present in human papilloma virus (HPV) infected cells. J Invest Dermatol 70:263−266

Ward M, Le Roux A, Small WP, Sircus W (1977) Malignant lymphoma and extensive viral wart formation in a patient with intestinal lymphangiectasia and lymphocyte depletion. Postgrad Med J 53:753−757

Wettstein FO, Stevens JG (1980) Distribution and state of viral nucleic acid in tumors induced by Shope papilloma virus. Cold Spring Harbor Conf Cell Prolif 7:301−307

Wettstein FO, Stevens JG (1981) Transcription of the viral genome in papillomas and carcinomas induced by the Shope virus. Virology 109:448−451

Wettstein FO, Stevens JG (1982) Variable-sized free episomes of Shope Papilloma virus DNA are present in all non-virus-producing neoplasms and integrated episomes are detected in some. Proc Natl Acad Sci USA 79:790−794

Woodruff JD, Braun L, Cavallieri R, Gupta P, Pass F, Shah KV (1980) Immunological identification of papillomavirus antigen in paraffin-processed condyloma tissues from the female genital tract. Obstet Gynecol 56:727−732

Yabe Y, Sadakane H, Isono H (1979) Connection between capsomeres in human papilloma virus. Virology 96:547−552

Yoshiike K, Defendi V (1977) Presence of deletion molecules in human wart virus DNA. J Virol 21:415−418

Zachow KR, Ostrow RS, Bender M, Watts S, Okagaki T,Pass F, Faras AJ (1982) Detection of human papillomavirus DNA in anogenital neoplasias. Nature 300:771—773

Zur Hausen H (1976) Condylomata acuminata and human genital cancer. Cancer Res 36:530

Zur Hausen H (1977a) Human papilloma viruses and their possible role in squamous cell carcinomas. Curr Top Microbiol Immunol 78:1—30

Zur Hausen H (1977b) Cell—virus gene balance hypothesis of carcinogenesis. Behring Inst Mitt 61:23—30

Zur Hausen H, Gissmann L, Steiner W, Dippold W, Dregger I (1975) Human papilloma virus and cancer. Bibl Haematologica 43:569—571

Rev. Physiol. Biochem. Pharmacol., Vol. 99
© by Springer-Verlag 1984

Peritubular Capillary, Interstitium, and Lymph of the Renal Cortex

G.G. PINTER* and K. GÄRTNER**

Dedicated to Professor Jan Brod of the
Medizinische Hochschule, Hannover, GFR,
on the occasion of his 70 birthay

Contents

---

* Recipient of the Alexander v. Humbolt Senior Award, 1980

  Department of Physiology, University of Maryland, School of Medicine, Baltimore, MD, USA

** Zentrales Tierlaboratorium, Medizinische Hochschule Hannover, GFR

1 Introduction

Research interest in the renal interstitium and renal lymph has increased in the past several years. Contributions using quantitative models attempted to add new findings to many general aspects of renal physiology and pathophysiology. The renewed interest in the renal interstitium and lymph is reflected by a recent editorial in the American Journal of Physiology (*Wolgast* et al. 1981), a chapter in a recent book on the Functional Ultrastructure of the Kidney (*Maunsbach* et al. 1980) and a state-of-the-art lecture at the 27th International Congress of Physiological Sciences (*Pinter* and *Wilson* 1981). This review will attempt to summarize recent research on the renal interstitium and lymph contributing to the general understanding of renal function in both health and disease.

2 Morphology and Function of Interstitium in the Kidney

This section emphasizes morphometric studies, primarily because in the past several years a convergence of experimental results has been achieved among various laboratories by use of careful fixation techniques. Moreover, some unresolved discrepancies remain in the evaluation of distribution space measurements. Measurements of distribution spaces have been reviewed recently by *Wolgast* et al. (1981).

2.1 Cortical Interstitium

Morphometric studies on the interstitial volume and interrelations between tubules and capillaries have been carried out in recent years by several groups of investigators. Most of the newer data have been obtained in the rat and tend to indicate that the volume fraction of the interstitium in this species is relatively small. According to *Pedersen* et al. (1980), the interstitium makes up 6.7% ± 0.2% by volume of the superficial cortex; 3.4% is wide interstitium, 0.7% narrow interstitium, and 2.6% interstitial cells. *Pfaller* and *Rittinger* (1977, 1980) found that in the outer cortex the extracellular space took up 4% and interstitial cells took up 5%, whereas in the juxtamedullary cortex these volumes were 3% and 4% respectively. Narrow interstitium was defined as that present between very closely apposed and largely parallel tubular and capillary walls, while wide interstitium was more irregular between walls at some distance from each other. *Langer* (1975) also distinguished between wide and narrow interstitial spaces, but gave no quantitative breakdown of these components. The figures given for the total interstitial volume agree with those of *Kriz* and *Napiwotzky* (1979) whose estimate, including interstitial cells, was approximately 9%. *Kriz* and *Napiwotzky* also estimated that 26% of the tubular surface is directly covered by capillaries without any interposed interstitial space.

Pedersen et al. (1980) made additional observations on the structure of the capillary wall facing the interstitium: they found that approximately two-thirds of the capillary wall facing narrow interstitium was of the fenestrated type, whereas only about one-third of the capillary wall facing wide interstitium was fenestrated.

The functional aspects of the specific features of the interstitium and the capillaries have remained largely unexplored. It appears as if the tubular reabsorbate is channeled into a structured space with preferential pathways for fluid flow; this could be important in view of the physiological control and some pathological conditions affecting the removal of reabsorbed fluid by capillary blood flow. *Vogel* et al. (1955), and *Lewy* and *Windhager* (1968) emphasized the role of peritubular oncotic and hydrostatic pressures on tubular reabsorption under normal conditions. In respect to the hydrostatic pressure in the renal interstitium, of particular interest are the experimental work and hypothesis of *Persson* et al. (1979) and *Persson* (1980), who showed that the sensitivity of the tubuloglomerular feedback mechanism is altered by manipulations which lead to changes in the interstitial hydrostatic pressure such that the sensitivity of the feedback mechanism appears to be enhanced when the interstitial pressure is low. This modulation of the feedback response could have a role in adaptation of the organism to unfavorable environmental conditions, e.g., water deprivation, when retention of fluid by means of suppressed glomerular filtration would favor the mechanisms which serve survival.

A largely unexplored area is the role of the interstitium in the pathomechanism of various forms of renal disease, when flow of tubular reabsorbate across the interstitium might be affected and, as a consequence, interstitial volume and pressure might be altered. The potential importance of the interstitium and lymph in glomerulonephritis and nephrotic syndrome has been discussed by *Rusznyák* et al. (1967). An important finding was reported by *Bohle* et al. (1977a, b), who observed a highly significant positive correlation between the renal cortical interstitial volume and the steady state plasma creatinine level in biopsy material obtained from patients suffering from membrane-proliferative glomerulonephritis. The volume of the cortical interstitium, measured with a morphometric method, was 8%–10% in subjects with minimal pathological change, whereas in severe disease it sometimes exceeded 40% of the total cortical volume. Work from the same laboratory (*Bohle* et al. 1979, 1981; *Riemenschneider* et al. 1980; *Mackensen-Haen* et al. 1981) supplemented this with

a finding of highly significant inverse correlation between cortical interstitial volume and creatinine clearance. However, they could discern no
correlation between the serum creatinine level and the morphological
status of the glomeruli. The observations of Bohle and associates point to
an as yet unexplored association between the renal cortical interstitium
and glomerular filtration.

2.2 Medullary Interstitium

Although attention here is focused primarily on the renal cortical interstitium, some recently gathered evidence on the renal medulla deserves to
be looked at. According to morphometric studies by *Pfaller* and *Rittinger*
(1977), the extracellular space including interstitial cells in the rat took up
7% and 11% respectively of the total volume in the outer and inner stripes
of the outer medulla. In the inner medulla the volume of the interstitium
increased from about 20% to near 30% from the outer to the inner layer,
with the interstitial cells amounted to approximately 10% throughout.

   Knepper et al. (1977) analyzed the renal medullary anatomy in rats
and rabbits. For technical reasons, the renal vein was tied off for up to
2 min before the kidney was removed. In the rat, a gradual increase of the
interstitial volume from the corticomedullary border to the papilla was
seen. Near the boundary between inner and outer medulla, the interstitium occupied approximately 10% of the total volume, in the papilla
nearly 30%. In the outer medulla the interstitial volume decreased toward
the cortex: in the outer stripe it was 4%–5%. In the rabbit a different
picture was seen: the interstitial volume was approximately 40% in the
papilla, but decreased precipitously within 3 mm of the papillary tip and
remained around 25% throughout the rest of both the inner and the
outer medulla. The results in the rat agree with those found by other investigators.

   Our own results (K. Gärtner and G.G. Pinter, unpublished work 1982)
indicated a large volume of distribution of extravascular albumin in the
rat renal medulla. Several aspects of the extravascular albumin pool in the
renal medulla are puzzling. Whereas extravasated macromolecules in the
cortical interstitium under normal conditions have access to the lymphatic
drainage (*Pinter* et al. 1981a), the route of macromolecules out of the
inner medullary interstitium is uncertain. According to morphological
evidence, the inner medulla has no lymphatics (*Kriz* and *Dieterich* 1970).
Recent investigations raise two interesting possibilities: (a) Studies by
Moffat (1969) and *Moffat* and *Williams* (1974) point to a potential
reentry of macromolecules from the interstitium into the vascular bed.
(b) *Reinking* and *Schmidt-Nielsen* (1981) observed that in rats and

hamsters, rhythmical contractions of the renal pelvic muscles exert intermittent pressure on the papilla and drive urine from the collecting ducts
into the pelvis. At the same time, these intermittent pressure waves initiate
a convective flow of fluid toward the corticomedullary border in the interstitium in longitudinal tissue spaces alongside the collecting ducts. These
preformed channels have no endothelial lining and were seen to pass
through interstitial cells. Bulk flow of interstitial fluid might thus provide
an exit for the extravasated macomolecules from the renal medulla. Many
details and implications of this possible mechanism have not yet been considered. Detailed studies by *Kriz* (1981) pointed out that the narrow interstitial spaces in the outer medulla are eminently suitable for countercurrent
exchange between adjacent structures. Convective flow of interstitial fluid
would presumably occur outside the vascular bundles and would not com-
promise countercurrent exchange.

   The renal medullary interstitium is rich in sulfated mucopolysaccharides, the functional significance of which remains hypothetical. The density of negative charges on these molecules was seen to be dependent on the
diuretic state (*Morard* and *Poirier* 1968; *Morard* and *Abadie* 1968). An
interaction between the mucopolysaccharides and plasma albumin, as described by *Laurent* and *Ogston* (1963) may occur in the medullary interstitium and may play a role in the urine-concentrating process (*Pinter*
1967).

3   Lymphatic System

3.1   Anatomy of Renal Lymphatic Vessels

A most careful study of the lymphatic system in ten mammalian species
was reported by *Kriz* and *Dieterich* (1970). According to these authors,
the cortical lymphatic capillaries originate, as a rule, in the loose connective
tissue around the interlobular blood vessels near the interlobular arterioles.
In the dog, there are both capsular and hilar routes for lymphatic drainage
from the kidney; in the rat a capsular drainage pathway can be seen only
very infrequently. Both hilar and – where existing – capsular lymphatic
vessels drain the cortical tissue, since, according to most investigators, the
medullary tissue does not contain lymphatics. *Albertine* and *O'Morchoe*
(1979, 1980) examined 100 vascular bundles and interbundle areas in 90
medullary tissue strips, and found that only one of these contained lymphatics. They also examined lymphatic vessels at the corticomedullary
border around the arcuate blood vessels. In 60 tissue blocks studies in
serial sections, only cortical lymphatic tributaries were seen. Morphological

evidence does not support the existence of lymphatic drainage of the renal medulla, and points to the need for a separate mechanism to provide for the drainage of extravasated macromolecules from the medullary interstitium (see above). *Atkins* et al. (1972) sought functional evidence for medullary contribution to the lymphatic drainage of the kidney. They determined in the dog the mean transit times of various tracers from arterial plasma simultaneously to the capsular and the hilar lymphatic drainage. They reasoned that if there were medullary drainage through the hilar lymph, because of the relatively slow flow of blood through the medulla some tracers should arrive with a delay in hilar lymph. No such delay was seen, and the authors concluded that the medulla makes no significant contribution to hilar lymph.

Recently, *Albertine* and *O'Morchoe* (1980, 1981) and *Yang* et al. (1981) studied the origin of renal lymphatics in the dog kidney and reviewed the literature on the mechanism of formation of renal lymph. They regularly encountered lymphatic capillaries within the lobules which were associated with tubules and glomeruli. However, the intralobular lymph capillaries were less numerous than the interlobular ones, and open intercellular gaps between endothelial cells of lymphatic capillaries were infrequent. *Albertine* and *O'Morchoe* (1980) suggested that a mechanism which calls for interstitial fluid entry into the lymph capillaries through open gaps between endothelial cells would not account for the rate of lymph formation in the kidney. They proposed a new hypothesis making vesicular transport of macromolecules through endothelial cells responsible for entry of proteins and fluid into the lymphatic vessels. Implicit in this suggestion is an assumption that in steady state there is a functional asymmetry between the luminal and abluminal surfaces of the endothelium of small lymphatic vessels. Such an asymmetry may be manifest in the rate of loading of vesicles or in the affinity of macromolecules to the membrane destined to become vesicular wall. No experimental demonstration of such an asymmetry is available at the present time (*Wagner* and *Casley-Smith* 1981).

3.2 Lymph

In general, two aspects of the lymph drainage of organs are considered to be of importance: (a) the contribution of lymph drainage to the fluid circulation and balance, and (b) the fact that lymph also constitutes a source of information about the interstitial fluid and, in turn, about the permeability of the capillaries to various substances, in particular macromolecules. These two points constitute the main areas of interest also for the lymph drainage of the kidney, where unique conditions prevail both for fluid circulation and lymphatic drainage.

3.2.1 Rate of Drainage

3.2.1.1 In the Dog

There are several, normally three to seven, hilar lymphatic vessels leaving the renal hilum near the main arterial branches. In addition to the hilar vessels there are also capsular lymphatic vessels which emerge on the renal surface. The capsular lymphatics follow a more or less contorted path in the direction of one of the poles, where they gather into larger lymphatic trunks. Lymphatic vessels do not tend to unite in a single collecting trunk and, for this reason, a determination of the total amount of lymph draining from the kidney by means of direct cannulation is not practicable. Various approaches have been tried to determine the total renal lymph drainage (see review by *O'Morchoe* and *O'Morchoe* 1968). All of the techniques used involved major surgical operations and a risk that both the blood circulation and lymph production of the kidney and the rest of the body could be compromised during the determinations.

Pinter et al. (1975b) devised a tracer method which did not interfere with the circulation, based on steady-state dilution of two distinguishable tracers of plasma albumin. One was infused into the artery of one kidney, the other into the renal venous blood, at equal volume rates. The arterial tracer reached the renal capillaries in a high concentration and a small fraction of it crossed the capillary walls into the interstitium and then into the renal lymph. However, over 99% of the arterial tracer passed through the kidney with the blood flow, and from this point circulated together with the venous tracer. By this use of tracers, one kidney was "isolated" from the rest of the body. From the excess amounts of arterial tracer in renal lymph and in thoracic duct lymph, the renal contribution to the thoracic duct lymph flow was calculated. The rate of average lymph production under control conditions was between 0.3 and 0.4 ml min^{-1} 100 g kidney^{-1}. The flow of lymph from one kidney amounted to approximately 20% of the total thoracic duct lymph flow. This figure indicated that in dogs anesthetized, immobilized, and deprived of food over a period of 24 h prior to the experiment, a large fraction of the thoracic duct lymph was contributed by the kidneys. *Pinter* et al. (1975b) ascertained that under normal conditions albumin entering the renal lymphatic vessels was quantitatively delivered into the thoracic duct. When the ureter was obstructed, the total lymph flow from the affected kidney was nearly doubled. Ureteric occlusion appeared to influence lymph production not only by the affected kidney, but also by other organs, as thoracic duct lymph flow increased in excess of the renal contribution.

3.2.1.2 In the Rat

Total renal lymph production in the rat was estimated by *Ulfendahl* et al. (1973) and *Atkins* et al. (1973). The method was based on the observation that a slight increase of renal venous pressure produces an instantaneous increase in the flow of lymph both from a cannulated renal lymphatic vessel and from the major abdominal lymph duct destined to become the thoracic duct after crossing the hiatus aorticus of the diaphragm. From the rates of simultaneous increase of lymph flows from the cannulated renal lymphatic vessel and from the large collecting abdominal lymph duct, the fraction of total renal contribution to the abdominal lymph flow was calculated. The method provided an estimate of about 3 μl min^{-1} g kidney^{-1}, agreeing with the figure found in the dog kidney. This method was used also by *Deen* et al. (1976), with similar results.

3.2.1.3 Potential Importance of Renal Lymphatic Drainage Under Pathological Conditions

Renal lymph flow in both dogs and rats amounts to 2–4 μl min^{-1} g kidney^{-1}, which compared with the plasma flow through the organ is not of sufficient magnitude to play an important role in the fluid balance of the kidney under normal physiological conditions. However, as quantitative data from *Vogel* et al. (1974) indicate, elevation of renal venous pressure leads to a large increase in renal lymph flow. Moreover, there are some pathological conditions in which renal lymph flow has been found to be much higher than normal. An increase of lymph flow in acute ureteric occlusion seen by *Pinter* et al. (1975b) has been mentioned above. As discussed by *Rusznyák* et al. (1967), alterations in renal lymph circulation were explored in chronic hydronephrosis. More recently, *Stork* et al. (1980) showed a large increase of the renal lymph flow in experimental diabetes in rats. Thus it appears that under certain pathological conditions, renal lymph can assume importance in the fluid balance of the kidney.

A significant amount of renin is transported from the kidney by renal lymph (*Lever* and *Peart* 1962; *Skinner* et al. 1963; *Bailie* et al. 1971; *Horkey* et al. 1971; *O'Morchoe* et al. 1981). Whether renal lymph has quantitatively important functions in the transport of various prostaglandins and the activated form of vitamin D, is not known at the present time.

3.2.2 Carrier of Information from the Interstitium

Uncertainties remain about the site and the mechanism of renal lymph formation and about the relationship between interstitial fluid and lymph.

Lymph formation is dependent on hydrostatic and oncotic pressure differences across blood and lymph capillary membranes, as postulated by *Starling* (1896). In the kidney the arterial and venous portions of the capillary bed are separated. In the glomeruli, hydrostatic and oncotic pressure conditions and the unique structure of the capillary wall ensure a high rate of filtration (*Brenner* et al. 1971, 1972; *Robertson* et al. 1972). In contrast, conditions in the peritubular capillary bed are suited for effective fluid reabsorption. The hydrostatic pressure in the peritubular capillaries was estimated to be approximately 10 mm Hg (*Källskog* et al. 1975). In the postglomerular vessels plasma proteins are more concentrated than in systemic blood, resulting in an inward-directed net (hydrostatic and oncotic) pressure difference of about 15 mm Hg at the beginning of the peritubular capillary. The driving force created by the oncotic pressure difference is dependent also on the reflexion coefficient to proteins of the capillary wall.

Although bulk flow of fluid takes place in only one direction (from the interstitium into the capillaries), diffusion through the peritubular capillary wall occurs in both directions. Small molecules diffuse rapidly: mannitol, creatinine and inulin are not present in the tubular reabsorbate as it enters the interstitium from the tubular epithelium, yet the concentration of these molecules in lymph is equal or nearly equal to that in renal venous blood. These substances gain access to renal lymph by diffusing outward from postglomerular blood vessels in a direction opposite to the convective flow of the tubular reabsorbate.

Under normal conditions, the glomerular filtrate contains very little protein (*Oken* and *Flamenbaum* 1971). The major part of filtered protein undergoes metabolic decomposition during reabsorption (*Maunsbach* 1976; *Bode* et al. 1980). Thus the main source of macromolecules in the interstitium and lymph is postglomerular plasma. These large molecules can pass across the capillary wall by various means – diffusion and convective transport through large pores, cytoplasmic vesicles, and various combinations of these mechanisms.

The permeability of the postglomerular capillaries and venules to large molecules may not be uniform along the entire surface of postglomerular blood vessels. The morphological observations of *Pedersen* et al. (1980) indicating differences in fenestration also imply differences in permeability characteristics.

3.2.2.1 Tubular Reabsorbate in Lymph

According to morphological evidence, the interstitial space in the renal cortical parenchyma, including both wide and narrow spaces between tubules and capillaries (see above), are continuous with the interlobular perivascular connective tissue, where the great majority of lymphatic capillaries originate (*Kriz* and *Dieterich* 1970). This continuity makes it possible, but does not necessarily ensure, that tubular reabsorbate become a constituent of renal lymph. Recently, *Cook* et al. (1982) demonstrated that reabsorbed glucose is present in renal lymph. The evidence indicated that the reabsorbed glucose reached the lymph by direct passage (diffusion and/or convection) through the interstitium. These experiments support the hypothesis that not only glucose but also other substances, including proteins, reach the lymph from those parts of the interstitium through which tubular reabsorbate flows. A similar conclusion was reached by Vogel and Gärtner (unpublished observation 1974).

3.2.2.2 Lymphatic Drainage of the Juxtaglomerular Region
As reviewed above, the presence of renin in renal lymph has been repeatedly confirmed. This finding could indicate a direct drainage into the lymph of the interstitial fluid from the vicinity of the glomeruli. *Rojo-Ortega* et al. (1973) demonstrated lymphatic capillaries near the vascular pole of glomeruli, whereas other authors (*Leiper* et al. 1977) could not find any such vessels. It should be noted that presence of renin in renal lymph does not constitute functional evidence for direct lymphatic drainage of the interstitium in the vicinity of the glomeruli, since renin can reach more remote lymphatic capillaries by means of diffusion and convection through the interstitium. In this connection, an important unanswered question is the drainage route of the glomerular mesangium. It has been suggested (*Tighe* 1975) that macromolecules and particulate substances, once released from the mesangium, are disposed of by the lymphatic system. *Leiper* et al. (1977) interpreted their experimental data as suggesting disposal through the macula densa region into the distal convoluted tubule.

3.2.2.3 Permeability of Peritubular Capillaries to Macromolecules
The mechanism by which constitutents of interstitial fluid enter the lymphatic circulation is not fully understood at the present time, and for this reason we cannot rule out the possibility that concentration differences exist between lymph and interstitial fluid. Generally, in using lymph as a source of information about the renal cortical interstitium the assumption is made that macromolecular concentrations are equal in the interstitial fluid and the lymph. This assumption, however, is not generally accepted, and the experimental findings of *Casley-Smith* and *Sims* (1976), *Casley-Smith* (1979), and *Witte* and *Zenzes-Geprägs* (1977) constitute examples where its validity is open to doubt. *Bell* et al. (1978) and *Pinter* et al. (1980) avoided this assumption, assuming instead that when a protein tracer is present at steady state, the *specific activities* are equal in lymph and interstitial fluid. Moreover, they assumed that in nonsteady state the specific activity of a given protein in lymph follows specific activity in the interstitium with measurable delay and dispersion.

In the following sections we consider various quantitative models and parameters pertaining to the permeability of peritubular capillary wall to macromolecules.

3.2.2.3.1   PS Product. A quantitative assessment of the permeability of the capillary wall to macromolecules requires information concerning the permeability times surface area (PS) product and reflexion coefficient (σ). Although a theoretical framework has long been available, such determinations have not been carried out until recently on the peritubular capillary bed in the renal cortex. A formula derived by *Perl* (1975) expresses the relationship between the concentration ratio of a macromolecule in lymph and in plasma to the PS product and solvent drag reflexion coefficient:

$$\frac{C_2}{C_1} = \frac{PS + 1/2\,(1-\sigma_f)\,L_2}{PS + 1/2\,(1+\sigma_f)\,L_2}$$

where C_2 and C_1 are respectively the interstitial and plasma concentrations of a macromolecules; P, S, and σ_f are weighted averages for permeability (cm s^{-1}), surface area (cm^2) of capillary bed in 100 g tissue, and solvent drag reflexion coefficients of the filtering and reabsorbing surfaces of the capillary wall, respectively; and L_2 is the volume flow of fluid (cm^3 s^{-1}) leaving the tissue which is "destined at least in part to become lymph" (*Perl* 1975; for further detail see original article). It is assumed that the lymph and interstitial concentrations of the macromolecule are equal.

According to this formula, experimental measurements of the PS product and the reflexion coefficient require the determination of lymph flow from the organ and the lymph concentration of the macromolecule under study. Experiments along this line were carried out by *Deen* et al. (1976), from whose data a control value of PS product of 10×10^{-4} ml s^{-1} 100 g cortex^{-1} can be calculated as an approximate estimate. *Bell* et al. (1978) based their experiments on the assumption of equality between interstitial and lymphatic specific activities of the labelled albumin. They argued that concentration, dilution, and random catabolism do not alter the specific activity of macromolecules during transport from the interstitium to the lymph: interstitial and lymph specific activity of a macromolecule should be equal under a wide variety of physiological and pathological conditions.

3.2.2.3.2 Ψ Clearance. The experiments by *Bell* et al. were based upon the central volume theorem (*Stephenson* 1948; *Meier* and *Zierler* 1954): The unidirectional clearance of plasma albumin from the capillaries to the interstitium (Ψ) was calculated from the mean transit time of albumin from arterial plasma to renal lymph ($\bar{t}$) and the extravascular distribution space of labeled albumin in the renal cortex (V_i) by the formula $\Psi = V_i/\bar{t}$. The mean transit time of albumin from arterial plasma to renal lymph was calculated by measuring the specific activities of tracer albumin in both arterial plasma and renal lymph over a period of approximately 90 min following an intravenous injection of labeled albumin. Using a non-compartmental model, the probability density function (PDF) of transit times was calculated by deconvolution and the mean of this PDF was taken as the mean transit time. The interstitial distribution volume of albumin in renal cortex was measured by determining simultaneously, with two distinguishable albumin tracers, the total and the intravascular distribution volumes of albumin. For the determination of the total distribution volume, one of the tracers was allowed to equilibrate over a period of 2 h or more; for the intravascular distribution volume, the second tracer was present in the circulation for a few minutes only. The difference between total and intravascular distribution volumes was taken as the measure of the interstitial distribution volume. When proteinuria was present (as for example in older and diabetic rats), the measurement was corrected for the presence of tracer albumin inside the tubules. In control animals $\bar{t}$ was near 40 min and V_i was approximately 1.7 ml 100 g tissue^{-1}, the latter figure corresponding to a physical volume occupied by the interstitium in the renal cortex of approximately 5%. The unidirectional clearance of albumin from the peritubular capillaries to the interstitium was calculated as 7×10^{-4} ml s^{-1} 100 g tissue^{-1}. It is of interest that the rate of albumin drainage from the kidney through the lymph is approximately 8×10^{-4} ml s^{-1} 100 g tissue^{-1} in normal rats. The good agreement between these figures indicates that under normal conditions, extravasated albumin from the renal cortical interstitium drains quantitatively through the lymph and no reentry of albumin into the peritubular capillaries takes place. (See Sect. 3.2.2.3.4 for additional discussion.)

3.2.2.3.3   Reflexion Coefficient. Pinter et al. (1975a) compared the PS product (i.e. net clearance) of albumin at the peritubular capillary wall to the unidirectional clearance and inferred that the reflexion coefficient of the wall of these capillaries to albumin is near unity. *Deen* et al. (1976) arrived at the same conclusion. *Bell* et al. (1978) assumed that along the capillary wall the reabsorbing surfaces are separate from the sites through which macromolecules are released into the interstitium, and using this model, calculated the solvent drag reflexion coefficient to be in excess of 0.99. Provided that the osmotic reflexion coefficient (σ_d) is equal to the solvent drag reflexion coefficient (σ_f), this figure indicates that the concentration difference of proteins across the reabsorbing surface of the capillary wall is fully effective in generating an osmotic pressure for fluid reabsorption into the capillary.

Different experimental approaches thus point to a high, nearly 1.0, solvent drag reflexion coefficient to albumin at the reabsorbing surface of the peritubular capillaries (*Bell* et al. 1978). In comparison to the reflexion coefficient of albumin at other capillary regions, this value appears to be high. It should be noted that the solvent drag reflexion coefficient expresses a relationship between measures of relative mobilities of solute and solvent across a membrane (cf. *Katchalsky* and *Curran* 1965, eq. 10–20, p. 122). The very high reflexion coefficient of the peritubular capillary wall to albumin indicates that (a) the membrane is relatively impermeable to albumin, and (b) there is a very large volume of convective flow of tubular reabsorbate through the membrane.

3.2.2.3.4 Functional Organization of the Interstitial Albumin Pool and a Scheme of Albumin Movement Within the Renal Cortex. Depending on the method, different experiments have provided differing – sometimes highly contrasting – characterizations of the interstitial albumin pool in the renal cortex. By using perfusion and washout of the renal circulation in dogs and rabbits, *Swann* (1960), *Ochwadt* (1964), and *Vogel* et al. (1969) concluded that the interstitial albumin pool amounted to the albumin content of about 4 ml blood plasma 100 g tissue^{-1}. The turnover rate of this pool was described as rapid (turnover time $\sim$2 min), and a substantial reentry of albumin from the interstitium into blood vessels was postulated. In contrast, *Bell* et al. (1978) and *Pinter* and *Wilson* (1981), using renal lymph in rats as a source of information, concluded that the cortical interstitial albumin pool represented an albumin content of 1–2 ml plasma 100 g cortex^{-1}. The turnover time of this pool was 30–40 min in the concentrating kidney, and under normal conditions virtually all of the interstitial albumin appeared to drain through the lymph; no reentry into vasculature from the interstitium was indicated by the data. Moreover, *Cook* et al. (1982) presented evidence that the pool studied through lymph is localized in the physical space in the cortical interstitium through which tubular reabsorbate passes.

These contrasting findings can be reconciled by the hypothesis that the rapidly exchanging pool seen with the washout method is different and separate from the pool which is discerned through lymph. The localization of the plasma exchange pool is uncertain; the possibility cannot be ruled out that a substantial part of it is intravascular in slowly perfused blood vessels (*Polosa* and *Hamilton* 1962). With respect to the interstitial pool draining through the lymph, the following scheme is postulated for the circulation of albumin: Through a relatively few specific sites (possibly large pores) on the capillary wall, macromolecules escape into the interstitium. We assume that these sites are functionally distinct, and possibly also morphologically separate from the reabsorbing surface of the peritubular capillary wall. At clustered protein releasing sites, convective flow may occur in a direction opposite to the prevailing inward flow of tubular reabsorbate. Normally, no macromolecules pass either by diffusion or by convection in either direction through the reabsorbing surface of the capillary wall. Thus under normal conditions the movement of macro-molecules from the capillaries to the interstitium is unidirectional and all albumin leaves the interstitium through lymphatic drainage.

This proposed mechanism is not compatible with a net outward bulk filtration from the peritubular capillaries. Even if all albumin were delivered into the peritubular interstitium by convection flow, the volume flow as-sociated with this movement of albumin under normal conditions would be approximately a thousand times *less* than the volume flow of tubular reabsorbate entering the peritubular capillaries.

3.2.2.4 A Model of Heterogeneity of the Interstitium
Wilson and *Pinter* (1979), using the experimental data of *Bell* et al. (1978), found that the single-compartment model of the renal cortical interstitium was not satisfactory in many experiments. The morphology of the renal cortical interstitium in the rat suggested that, whereas small regions in the interstitium might function as individual well-mixed compartments, the entire interstitium corresponds to a large collection of such individual units. The turnover rates of the individual units were assumed to be statistically distributed according to a two-parameter gamma PDF. In each experiment, this model provided a highly improved fit to the experimental data. Using this model, the PDF of turnover rates, its mean, and its variance were derived for each experiment. The variance allowed a conclusion about the degree of heterogeneity of the interstitial space. Populations of normal control and diabetic rats were compared, and the degree of heterogeneity was seen to be consistently higher in severely diabetic animals (*Wilson* and *Pinter* 1979).

4 Diabetes Mellitus and Other Experimental and Pathological Conditions

4.1 The Interstitium and Lymph in Experimental Diabetes Mellitus

Stork et al. (1980) studied the interstitial distribution volume of labeled albumin and the permeability of the peritubular capillaries in rats made diabetic with streptozotocin. The experiments were carried out at various times after the injection of the diabetogenic agent, and the experimental animals were classified into two groups according to the severity of the disease. Lymph flow increased to several times the normal rate in severely diabetic animals. The extravascular distribution volume of albumin in the renal cortex increased and the mean transit time of labeled albumin from arterial plasma to renal lymph slightly decreased, indicating a much increased unidirectional clearance of albumin across the interstitial space. These experimental results raise the possibility that an increase in peritubular capillary permeability to albumin also occurs in diabetic nephropathy in human subjects. *Pinter* et al. (to be published) suggested that simultaneous injury to both glomerular and peritubular capillary regions in the kidney can lead to a rapid loss of renal function, as the functional consequence of these injuries tend toward mutual reinforcement, resulting in a vicious circle.

The experiments of *Stork* et al. (1980) supported the notion that the pathological process in experimental diabetes produces increased structural and functional heterogeneity of the capillary permeability and the interstitium of the renal cortex. A model which allows a conclusion about the degree of heterogeneity of the interstitial albumin pool was introduced by *Wilson* and *Pinter* (1979) (see above).

4.2 Other Experimental and Pathological Conditions

The interstitial distribution volume of albumin in the renal cortex, the mean transit time of tracer albumin from peritubular capillary plasma to lymph, and the unidirectional clearance, calculated from these two measurements, did not differ from control values in the surgically denervated rat kidney (*Stork* et al. 1977) or in a $HgCl_2$ model of acute renal failure (*Pinter* et al. 1981b).

5 Future Prospects for Research on Renal Interstitium and Lymph

Until the end of the 1950's, research interest in lymph in general, and renal lymph, in particular, was intensive as shown by the comprehensive review article of *Mayerson* (1963) in the circulation section of the *Handbook of Physiology*. Since the 1960's, research activity on renal lymph has declined as renal research has received strong directions from the many successful applications of various microtechniques. These microtechniques are excellently suited to studying the functions of component elements of the kidney, and undoubtedly this very fruitful trend will continue in the future, but nonetheless, the current increase of interest in research on the renal interstitium and lymph does not appear to be accidental. Studies on specific functions of component elements are bound to raise questions about the means of integration, i.e., how global kidney function emerges from the individual details.

Studies on integrated functions usually begin with inquiries into inter-actions and coordinations between component elements: in the case of the kidney, how various building blocks, among them nephrons and vascular elements, tend to work together. The interstitium plays a prominent role in providing a functional connecting pathway between individual component elements, and, as discussed above, lymph is a ready carrier of information from the interstitium. Thus research interest in the renal interstitium and lymph should increase with interest in the integrative aspects of renal physiology.

Research on renal interstitium and lymph promises to be particularly rewarding in studies on renal disease. In some recent instances exceedingly large pathological changes have been seen in interstitium and lymph (*Bohle* et al. 1977; *Stork* et al. 1980), and the exploration of the functional significance of these changes also promises to be fruitful (*Pinter* et al., to be published). The observation of large changes points toward sensitivity of the interstitium to pathological influences. Further research on the renal interstitium and lymph will provide important information for physiologists, pathophysiologists, and clinicians who are striving to understand the integrated mechanisms of renal function in both health and disease.

Acknowledgements. The major part of the authors' experimental work referred to in this review was carried out with the support of USPHS grant AM 17093, at the Department of Physiology, University of Maryland, School of Medicine, and the Deutsche Forschungsgemeinschaft grant SFB 146/B 9, at the Medizinische Hochschule, Hannover. The authors gratefully acknowledge the generous assistance of the Alexander von Humbolt Foundation of the Federal Republic of Germany, which provided means for their cooperation.

References

Albertine KH, O'Morchoe CCC (1979) Distribution and density of the canine renal cortical lymphatic system. Kidney Int 16:470–480
Albertine KH, O'Morchoe CCC (1980) Renal lymphatic ultrastructure and translymphatic transport. Microvasc Res 19:338–351
Albertine KH, O'Morchoe CCC (1981) An ultrastructural study of the transport pathways across arcuate, interlobar, hilar and capsular lymphatics in the dog kidney. Microvasc Res 21:351–361
Atkins JL, O'Morchoe CCC, Pinter GG (1972) Simultaneous studies of hilar and capsular renal lymph. Life Sci 11(I):1007–1010
Atkins JL, O'Morchoe CCC, Ulfendahl HR, Wolgast M, Agerup B, Pinter GG (1973) Total lymph flow in dogs and rats. ASN 6th Annual Meeting (Abstracts), p 5
Bailie MD, Rector FC Jr, Seldin DW (1971) Angiotensin II in arterial and renal venous plasma and renal lymph in the dog. J Clin Invest 50:119–126
Bell DR, Pinter GG, Wilson PD (1978) Albumin permeability of the peritubular capillaries in rat renal cortex. J Physiol (Lond) 279:621–640

Bode F, Ottosen PD, Madsen KM, Maunsbach AB (1980) Does transtubular transport of intact protein occur in the kidney? In: Maunsbach AB, Olsen TS, Christensen ET (eds) Functional ultrastructure of the kidney. Academic Press, New York, pp 385–395

Bohle A, Glomb D, Grund KE, Mackensen S (1977a) Correlations between relative interstitial volume of the renal cortex and serum creatinine concentration in minimal changes with nephrotic syndrome and in focal sclerosing glomerulonephritis. Virchows Arch [Pathol Anat] 376:221–232

Bohle A, Grund KE, Mackensen S, Tolon M (1977b) Correlations between renal interstitium and level of serum creatinine. Virchows Arch [Pathol Anat] 373:15–22

Bohle A, Christ H, Grund KE, Mackensen S (1979) The role of the interstitium of the renal cortex in renal disease. Contrib Nephrol 16:109–114

Bohle A, Gise HV, Mackensen-Haen S, Stark-Jakob B (1981) The obliteration of the postglomerular capillaries and its influence upon the function of both glomeruli and tubuli. Klin Wochenschr 59:1043–1051

Brenner BM, Troy JL, Daugharty TM (1971) The dynamics of glomerular ultrafiltration in the rat. J Clin Invest 50:1776–1780

Brenner BM, Troy JL, Daugherty TM, Deen WM, Robertson CR (1972) Dynamics of glomerular ultrafiltration in the rat. II. Plasma-flow dependence of GFR. Am J Physiol 223:1184–1190

Casley-Smith JR (1979) Fine structural study of variations in protein concentration in lacteals during compression and relaxation. Lymphology 12:59–65

Casley-Smith JR, Sims MA (1976) Protein concentrations in regions with fenestrated and continuous blood capillaries and in initial and collecting lymphatics. Microvasc Res 12:245–257

Cook VL, Reese AH, Wilson PD, Pinter GG (1982) Access of reabsorbed glucose to renal lymph. Experientia 38:108–109

Deen WM, Ueki JF, Brenner BM (1976) Permeability of renal peritubular capillaries to neutral dextrans and endogenous albumin. Am J Physiol 231:283–291

Gärtner K, Vogel G, Ulbrich M (1968) Untersuchungen zur Penetration von Makromolekülen (Polyvinylpirrolidon) durch glomeruläre und postglomeruläre Kapillaren in der Harn- und Nierenlymphe und zur Größe der extravasalen Umwälzung von I-131 Albumin im Interstitium der Niere. Pfluegers Arch 298:305–321

Horkey K, Rojo-Ortega JM, Rodriguez J, Boucher R, Genest J (1971) Renin, renin substrate, and angiotensin I-converting enzyme in lymph of rats. Am J Physiol 220:307–331

Källskog O, Lindbom LO, Ulfendahl HR, Wolgast M (1975) Kinetics of the glomerular ultrafiltration in the rat kidney. An experimental study. Acta Physiol Scand 95:293–300

Katchalsky A, Curran PG (1965) Nonequilibrium thermodynamics in biophysics. Harvard Univ Press, Cambridge, Mass

Knepper MA, Danielson RA, Saidel GM, Post RS (1977) Quantitative analysis of renal medullary anatomy in rats and rabbits. Kidney Int 12:313–323

Kriz W (1981) Structural organization of the renal medulla: comparative and functional aspects. Editorial review. Am J Physiol 241:R3–R16

Kriz W, Dieterich HJ (1970) Das Lymphgefäßsystem der Niere bei einigen Säugetieren. Licht- und elektronenmikroskopische Untersuchungen. Anat Embryol (Berl) 131:111–147

Kriz W, Napiwotzky P (1979) Structural and functional aspects of the renal interstitium. Contrib Nephrol 16:104–108

Langer KH (1975) Nitereninterstitium – Feinstrukturen und kapillare Permeabilität. I. Feinstrukturen der zellulären und extrazellulären Komponenten des peritubulären Niereninterstitiums. II. Elektronenmikroskopische Permeabilitätsstudien an peritubulären Kapillaren der Niere. III. Untersuchungen über die Verteilung von Tracer-Proteinen im peritubulären Interstitium und tubulären Labyrinth. Cytobiologie 10:161–216

Laurent TC, Ogston AG (1963) The interaction between polysaccharides and other macromolecules. 4. The osmotic pressure of mixtures of serum albumin and hyaluronic acid. Biochem J 89:249–253

Leiper JM, Thomson D, MacDonald MK (1977) Uptake and transport of imposil by the glomerular mesangium in the mouse. Lab Invest 37:526–533

Lever AF, Peart WS (1962) Renin and angiotensin-like activity in renal lymph. J Physiol 160:548–563

Lewy JE, Windhager E (1968) Peritubular control of proximal tubular reabsorption in the rat kidney. Am J Physiol 214:943–954

Mackensen-Haen S, Bader R, Grund KE, Bohle A (1981) Correlations between renal cortical interstitial fibrosis, atrophy of the proximal tubules and impairment of the glomerular filtration rate. Clin Nephrol 15:167–171

Maunsbach AB (1976) Cellular mechanisms of tubular protein transport. In: Thurau K (ed) Kidney and urinary tract physiology II, vol 2. University Park Press, Baltimore, pp 147–167. Guyton AC (ed) Int Rev Physiol Series

Maunsbach AB, Olson TS, Christensen L (1980) Functional ultrastructure of the kidney. Academic Press, New York, pp 399–474

Mayerson HS (1963) The physiologic importance of lymph. In: Hamilton WF (ed) Circulation, vol II. Am Physiol Soc, Washington, DC (Handbook of physiology, sect 2, pp 1035–1073)

Meier P, Zierler KL (1954) On the theory of the indicator-distribution method for measurements of blood flow and volume. J Appl Physiol 6:731–744

Moffat DB (1969) Extravascular protein in the renal medulla. Q J Exp Physiol 54: 60–67

Moffat DB, Williams MMM (1974) The effect of antidiuretic hormone on the extravascular protein in the renal medulla. Experientia 15:556–557

Morard JC, Poirier MF (1968) Fonction des mucopolysaccharides et des mucoides acides de la medullaire rénale dans l'elaboration de l'urine. I. Etudes histochimiques au cours de la diurese normale. J Physiol (Paris) 60:297–321

Morard JC, Abadie A (1968) Fonction des mucopolysaccharides et des mucoides acides de la medullaire rénale dans l'elaboration de l'urine. II. Etudes histochimiques au cours de diurese controlées. J Physiol (Paris) 60:323–356

Ochwadt B (1964) The measurement of intrarenal blood flow distribution by wash-out technique. Int Congr Series 78:62–64

Oken DE, Flamenbaum W (1971) Micropuncture studies of proximal tubule albumin concentrations in normal and nephrotic rats. J Clin Invest 50:1498–1505

O'Morchoe CCC, O'Morchoe PJ (1968) Renal contribution to thoracic duct lymph in dogs. J Physiol (Lond) 194:305–315

O'Morchoe CCC, O'Morchoe PJ, Albertine KH, Jarosz HM (1981) Concentration of renin in the renal interstitium, as reflected in lymph. In: Berlyne GM, Thomas S (eds) Renal physiology, vol 4. Karger, Basel, pp 199–206

Pedersen JC, Persson AEG, Maunsbach AB (1980) Ultrastructure and quantitative characterization of the cortical interstitium in the rat kidney. In: Maunsbach AB, Olsen TS, Christensen EL (eds) Functional ultrastructure of the kidney. Academic Press, London New York, pp 443–457

Perl W (1975) Convection and permeation of albumin between plasma and interstitium. Microvasc Res 10:83–94

Persson AEG (1980) Functional aspects of the renal interstitium. In: Maunsbach AB, Olsen TS, Christensen EL (eds) Ultrastructure of the kidney. Academic Press, London New York, pp 399–410

Persson AEG, Muller-Suur R, Selen G (1979) Capillary oncotic pressure as a modifier for tubuloglomerular feedback. Am J Physiol 236:F97–F102

Pfaller W, Rittinger M (1977) Quantitative Morphologie der Niere. Mikroskopie 33: 74–79

Pfaller W, Rittinger M (1980) Quantitative morphlogy of the rat kidney. Int J Biochem 12:17–22

Pinter GG (1967) Distribution of chylomicrons and albumin in dog kidney. J Physiol (Lond) 192:761–772

Pinter GG, Wilson PD (1981) Renal cortical interstitium and renal lymph with remarks on a stochastic conception of the reflexion coefficient of the peritubular capillary wall. In: Takacs L (ed) Advances in Physiological Science, vol II. Pergamon Press, New York, pp 57–73

Pinter GG, Atkins JL, Bell DR, Stork JE (1975a) Permeability of the peritubular capillaries to albumin. Implication for the reflexion coefficient. (Abstract) 6th International Congress of Nephrology

Pinter GG, O'Morchoe CCC, Atkins JL (1975b) Quantitative measurement of total renal lymph drainage. An experimental tracer study in dogs. In: Winkel K zum, Blaufox MD, Funck-Brentano JL (eds) Radionuclides in nephrology). Thieme, Stuttgart, pp 29–33

Pinter GG, Wilson PD, Bell DR, Atkins JL, Stork JE (1980) Interstitial albumin pool in the renal cortex: its turnover and the permeability of peritubular capillaries. In: Maunsbach AB, Olsen TS, Christensen EI (eds) Functional ultrastructure of the kidney. Academic Press, London New York, pp 411–422

Pinter GG, Alt J, Gärtner K, Lübow J, Stolte H, Wilson PD (1981a) Do extravasated protein molecules reenter the peritubular capillaires in the renal cortex of rats. (Abstracts) 8th International Congress of Nephrology

Pinter GG, Reese DA, Gärtner K, Reese AH (1981b) Lack of short-term effect of $HgCl_2$ on the albumin permeability of the peritubular capillaries in rat renal cortex (Abstract). Tel Aviv Satellite Symposium on Acute Renal Failure, Tel Aviv, Israel, p 54

Pinter GG, Wilson PD, Stork JE, Fajer AB (to be published) Functional manifestations of microangiopathy in experimental diabetes mellitus in the renal postglomerular circulation. Proceedings of the workshop Lessons from Animals Diabetes, Jerusalem, November 1982

Polosa C, Hamilton WF (1962) The relation between cells and plasma within the renal vasculature. Arch Int Pharmacodyn 140:294–307

Reinking LN, Schmidt-Nielsen B (1981) Peristaltic flow of urine in the renal papillary collecting ducts of hamsters. Kidney Int 20:55–60

Riemenschneider T, Mackensen-Haen S, Christ H, Bohle A (1980) Correlation between endogenous creatinine clearance and relative interstitial volume of the renal cortex in patients with diffuse membranous glomerulonephritis having a normal serum creatinine concentration. Lab Invest 43:145–149

Robertson CR, Deen WM, Troy JL, Brenner BM (1972) Dynamics of glomerular ultra-filtration in the rat. III. Hemodynamics and autoregulation. Am J Physiol 223: 1191–1200

Rojo-Ortega JM, Yeghiayan E, Genest J (1973) Lymphatic capillaries in the renal cortex of the rat. Lab Invest 29:336–341

Rusznyák I, Földi M, Szabo G (1967) Lymphatics and lymph circulation. Physiology and pathology. Pergamon Press, New York

Skinner SL, McCubbin JW, Page IH (1963) Angiotensin in blood and lymph following reduction in renal arterial perfusion pressure in dogs. Circ Res 13:336–345

Starling EH (1896) On the absorption of fluids from the connective tissue spaces. J Physiol (Lond) 19:312–326

Stephenson JL (1948) Theory of the measurement of blood flow by the dilution of an indicator. Bull Math Biophys 10:117–121

Stork JE, Wilson PD, Reese DA, Urbaitis BK, Blake WD, Pinter GG (1977) Lack of effect of splanchnic nerve sectioning on the albumin permeability of the peritubular capillaries. (Abstract) Proceedings of 2nd European Colloquium on Renal Physiology

Stork JE, Wilson PD, Pinter GG (1980) Interstitital albumin pool in renal cortex: The permeability of the peritubular capillaries in experimental diabetes mellitus. In: Maunsbach AB, Olsen TS, Christensen EI (eds) Functional ultrastructure of the kidney. Academic Press, London New York, pp 423–429

Swann HG (1960) The functional distension of the kidney: a review. Tex Rep Biol Med 18:566–595

Tighe JR (1975) The mesangium in glomerular disease. Proc R Soc Med 68:151–158

Ulfendahl HR, Pinter GG, Atkins JL, Wolgast M, Agerup B (1973) Total lymph flow of rat kidney. Acta Physiol Scand [Suppl] 396:92

Vogel G, Heym E, Andersohn K (1955) Versuche zur Bedeutung kolloidosmotischer Druckdifferenzen für einen passiven Transportmechanismus in den Nierenkanälchen. Anat Embryol (Berl) 126:485–489

Vogel G, Ulbrich M, Gärtner K (1969) Über den Austausch des extravasalen Plasma-Albumins (^{131}J-Albumin) der Niere mit dem Blut und den Abfluß von Makromolekülen (Polyvinylpyrrolidon) mit der Nierenlymphe bei normaler und durch Furosemid gehemmter tubulärer Reabsorption. Pfluegers Arch 305:47–64

Vogel G, Gärtner K, Ulbrich M (1974) The flow rate and macromolecule content of hilar lymph from the rabbit's kidney under conditions of renal venous pressure elevation and restriction of renal function – studies on the origin of renal lymph. Lymphology 3:136–143

Wagner RC, Casely-Smith JR (1981) Endothelial vesicles (review). Microvasc Res 21:267–298

Wilson PD, Pinter GG (1979) A model for heterogeneity of kinetics of albumin transport in the renal cortical interstitium. Math Biosci 46:1–10

Witte S, Zenzes-Geprägs S (1977) Extravascular protein measurements in vivo and in sites by ultramicrospectrophotometry. Microvasc Res 13:225–231

Wolgast M, Larson M, Nygren K (1981) Functional characteristic of the renal interstitium. Am J Physiol 241:F105–F111 (Editorial review)

Yang VV, O'Morchoe PJ, O'Morchoe CCC (1981) Transport of protein across lymphatic endothelium in the rat kidney. Microvasc Res 21:75–91

Author Index

Page numbers in *italics* refer to the bibliography.

Subject Index

Of Further Interest from this Series

Volume 91

1981. 18 figures. V, 227 pages
ISBN 3-540-10961-7

Contents: H. Favre, N.S. Bricker: The Pathology of Marginal Renal Function. - G. Hocman: Human Thyroxine Binding Globulin (TBG). - H.M. Verheij, A.J. Slotboom, G.H. de Haas: Structure and Function of Phospholipase A₂.

Volume 92

1982. 13 figures. V, 223 pages
ISBN 3-540-11105-0

Contents: P. Honerjäger: Cardioactive Substances that Prolong the Open State of Sodium Channels. - F.G. Knox, J.A. Haas: Factors Influencing Renal Sodium Reabsorption in Volume Expansion. - J. Comsa, H. Leonhardt, H. Wekerle: Hormonal Coordination of the Immune Response.

.

Volume 93

1982. 22 figures. V, 213 pages
ISBN 3-540-11297-9

Contents: H.-H. Wellhöner: Tetanus Neurotoxin. - R.D. Manning, Jr., A.C. Guyton: Control of Blood Volume. - P.C. Heinrich: Proteolytic Processing of Polypeptides During the Biosynthesis of Subcellular Structures.

Springer-Verlag
Berlin
Heidelberg
New York
Tokyo

Volume 94

1982. 52 figures. V, 207 pages
ISBN 3-540-11701-6

Contents: M. Vogt: Obituary Joshŭa Harold Burn (1892–1981). - A. Malliani: Cardiovascular Sympathetic Afferent Fibers. - G.F. DiBona: The Functions of the Renal Nerves.

Volume 95

1983. 50 figures. V, 223 pages
ISBN 3-540-11736-9

Contents: V.B. Brooks: Study of Brain Function by Local, Reversible Cooling. - H. Rasmussen, D.M. Waisman: Modulation of Cell Function in the Calcium Messenger System. - K.E.O. Åkerman, D.G. Nicholls: Physiological and Bioenergetic Aspects of Mitochondrial Calcium Transport.

Volume 96

1983. 21 figures. V, 194 pages
ISBN 3-540-11849-7

Contents: H. Murer, G. Burckhardt: Membrane Transport of Anions Across Epithelia of Mammalian Small Intestine and Kidney Proximal Tubule. - W. Osswald, S. Guimarães: Adrenergic Mechanisms in Blood Vessels: Morphological and Pharmacological Aspects. - O.H. Wieland: The Mammalian Pyruvate Dehydrogenase Complex: Structure and Regulation.

Volume 97

1983. 15 figures. V, 176 pages
ISBN 3-540-12135-8

Contents: P.R. Stanfield: Tetraethylammonium Ions and the Potassium Permeability of Excitable Cells. - G.A. Quamme, J.H. Dirks: Renal Magnesium Transport. - R. Fänge: Gas Exchange in Fish Swim Bladder.

Volume 98

1983. 94 figures. V, 255 pages
ISBN 3-540-12817-4

Contents: H. Blaschko: Obituary Hans Adolf Krebs 1900–1981. - J. Szentágothai: The Modular Architectonic Principle of Neural Centers. - S.D. Erulkar: The Modulation of Neurotransmitter Release at Synaptic Junctions. - J. Krüger: Simultaneous Individual Recordings From Many Cerebral Neurons: Techniques and Results.

Current Topics in Microbiology and Immunology

Editors: M.Cooper, W.Henle, P.H.Hofschneider,
H.Koprowski, F.Melchers, R.Rott,
H.G.Schweiger, P.K.Vogt, R.Zinkernagel

Volume 98
Retrovirus Genes in Lymphocyte Function and Growth

Editors: E.Wecker, I.Horak
1982. 8 figures. VIII, 142 pages
ISBN 3-540-11225-1

Volume 100
T Cell Hybridomas

A Workshop at the Basel Institute for Immunology
Organized and edited by H.V.Boehmer, W.Haas,
G.Köhler, F.Melchers, J.Zeuthen
With the collaboration of S.Buser-Boyd
1982. 52 figures. XI, 262 pages
ISBN 3-540-11535-8

Volume 101
Tumorviruses, Neoplastic Transformation and Differentiation

Editors: T.Graf, R.Jaenisch
1982. 27 figures. VIII, 198 pages
ISBN 3-540-11665-6

Volume 103
Retroviruses 1

Editors: P.K.Vogt, H.Koprowski
1983. 16 figures. V, 146 pages
ISBN 3-540-12167-6

Volume 104
New Developments in Diagnostic Virology

Editor: P.A.Bachmann
1983. 117 figures. XII, 330 pages
ISBN 3-540-12171-4

Volume 105
1983. 35 figures. III, 184 pages
ISBN 3-540-12492-6

Contents: E.G.Strauss, J.H.Strauss: Replication Strategies of the Single Stranded RNA Viruses of Eukaryotes. – P.L.Marion, W.S.Robinson: Hepadna Viruses: Hepatitis B and Related Viruses. – M.K.Estes, E.L.Palmer, J.F.Obijeski: Rotaviruses: A Review.

Volume 106
Mouse Mammary Tumor Virus

Editors: P.K.Vogt, H.Koprowski
1983. 12 figures. VIII, 104 pages
ISBN 3-540-12828-X

Volume 107
Retroviruses 2

Editors: P.K.Vogt, H.Koprowski
1984. 26 figures. Approx. 190 pages
ISBN 3-540-12384-9

Volume 108
Methylation of DNA

Editor: T.A.Trautner
1984. 22 figures. Approx. 180 pages
ISBN 3-540-12849-2

Springer-Verlag
Berlin
Heidelberg
New York
Tokyo